Communications
in Computer and Information Science   1482

More information about this series at http://www.springer.com/series/7899

Gang Wang · Arridhana Ciptadi ·
Ali Ahmadzadeh (Eds.)

Deployable Machine Learning for Security Defense

Second International Workshop, MLHat 2021
Virtual Event, August 15, 2021
Proceedings

 Springer

Editors
Gang Wang ⓘ
University of Illinois at Urbana-Champaign
Urbana, IL, USA

Arridhana Ciptadi
Truera Inc.
Redwood City, CA, USA

Ali Ahmadzadeh
Blue Hexagon Inc.
Sunnyvale, CA, USA

ISSN 1865-0929   ISSN 1865-0937   (electronic)
Communications in Computer and Information Science
ISBN 978-3-030-87838-2   ISBN 978-3-030-87839-9   (eBook)
https://doi.org/10.1007/978-3-030-87839-9

This Springer imprint is published by the registered company Springer Nature Switzerland AG
The registered company address is: Gewerbestrasse 11, 6330 Cham, Switzerland

Preface

In recent years, we have seen machine learning algorithms, particularly deep learning algorithms, revolutionizing many domains such as computer vision, speech, and natural language processing. In contrast, the impact of these new advances in machine learning is still limited in the domain of security defense. While there is research progress in applying machine learning for threat forensics, malware analysis, intrusion detection, and vulnerability discovery, there are still grand challenges to be addressed before a machine learning system can be deployed and operated in practice as a critical component of cyber defense. Major challenges include, but are not limited to, the scale of the problem (billions of known attacks), adaptability (hundreds of millions of new attacks every year), inference speed and efficiency (compute resource is constrained), adversarial attacks (highly motivated evasion, poisoning, and trojaning attacks), the surging demand for explainability (for threat investigation), and the need for integrating humans (e.g., SOC analysts) in the loop.

To address these challenges, we hosted the second International Workshop on Deployable Machine Learning for Security Defense (MLHat 2021). The workshop was co-located with 26th ACM SIGKDD Conference on Knowledge Discovery and Data Mining (KDD 2021). This workshop brought together academic researchers and industry practitioners to discuss the open challenges, potential solutions, and best practices to deploy machine learning at scale for security defense. The goal was to define new machine learning paradigms under various security application contexts and identify exciting new future research directions. The workshop had a strong industry presence to provide insights into the challenges in deploying and maintaining machine learning models and the much-needed discussion on the capabilities that state-of-the-art systems have failed to provide.

The workshop received seven complete submissions as "novel research papers". All of the submissions were single-blind. Each submission received three reviews from the Technical Program Committee members. In total, six full papers were selected and presented during the workshop.

August 2021

Gang Wang
Arridhana Ciptadi
Ali Ahmadzadeh

Organization

Organizing and Program Committee Chairs

Gang Wang   University of Illinois at Urbana-Champaign, USA
Arridhana Ciptadi   Truera, USA
Ali Ahmadzadeh   Blue Hexagon, USA

Program Committee

Sadia Afroz   Avast, USA
Siddharth Bhatia   National University of Singapore, Singapore
Wenbo Guo   Pennsylvania State University, USA
Zhou Li   UC Irvine, USA
Fabio Pierazzi   King's College London, UK
Alborz Rezazadeh   LG AI Research Lab, Canada
Gianluca Stringhini   Boston University, USA
Binghui Wang   Duke University, USA
Ting Wang   Pennsylvania State University, USA

Contents

Machine Learning for Security

STAN: Synthetic Network Traffic Generation with Generative Neural Models

Shengzhe Xu[1]([✉]), Manish Marwah[2], Martin Arlitt[2], and Naren Ramakrishnan[1]

[1] Department of Computer Science, Virginia Tech, Arlington, VA, USA
shengzx@vt.edu, naren@cs.vt.edu
[2] Micro Focus, Santa Clara, CA, USA
{manish.marwah,martin.arlitt}@microfocus.com

Abstract. Deep learning models have achieved great success in recent years but progress in some domains like cybersecurity is stymied due to a paucity of realistic datasets. Organizations are reluctant to share such data, even internally, due to privacy reasons. An alternative is to use synthetically generated data but existing methods are limited in their ability to capture complex dependency structures, between attributes and across time. This paper presents *STAN* (Synthetic network Traffic generation with Autoregressive Neural models), a tool to generate realistic synthetic network traffic datasets for subsequent downstream applications. Our novel neural architecture captures both temporal dependencies and dependence between attributes at any given time. It integrates convolutional neural layers with mixture density neural layers and softmax layers, and models both continuous and discrete variables. We evaluate the performance of *STAN* in terms of quality of data generated, by training it on both a simulated dataset and a real network traffic data set. Finally, to answer the question—can real network traffic data be substituted with synthetic data to train models of comparable accuracy?—we train two anomaly detection models based on self-supervision. The results show only a small decline in accuracy of models trained solely on synthetic data. While current results are encouraging in terms of quality of data generated and absence of any obvious data leakage from training data, in the future we plan to further validate this fact by conducting privacy attacks on the generated data. Other future work includes validating capture of long term dependencies and making model training more efficient.

1 Introduction

Cybersecurity has become a key concern for both private and public organizations, given the prevalence of cyber-threats and attacks. In fact, malicious cyber-activity cost the U.S. economy between $57 billion and $109 billion in 2016 [33], and worldwide yearly spending on cybersecurity reached $1.5 trillion in 2018 [29].

© Springer Nature Switzerland AG 2021
G. Wang et al. (Eds.): MLHat 2021, CCIS 1482, pp. 3–29, 2021.
https://doi.org/10.1007/978-3-030-87839-9_1

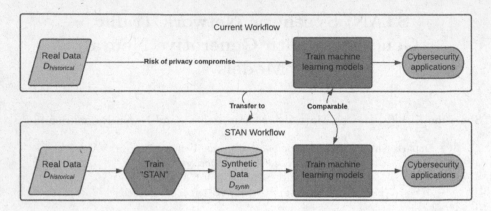

Fig. 1. *STAN* The top figure shows a simplified workflow where real data ($D_{historic}$) is used to train a machine learning model for cybersecurity applications; however, use of real data may result in a privacy compromise. The bottom figure shows the proposed workflow, where machine learning models are trained using realistic synthetic data (D_{synth}) generated by STAN.

To gain insights into and to counter cybersecurity threats, organizations need to sift through large amounts of network, host, and application data typically produced in an organization. Manual inspection of such data by security analysts to discover attacks is impractical due to its sheer volume, e.g., even a medium-sized organization can produce terabytes of network traffic data in a few hours. Automating the process through use of machine learning tools is the only viable alternative. Recently, deep learning models have been successfully used for cybersecurity applications [5,16], and given the large quantities of available data, deep learning methods appear to be a good fit.

However, although large amounts of data is available to security teams for cybersecurity machine learning applications, it is sensitive in nature and access to it can result in privacy violations, e.g., network traffic logs can reveal web browsing behavior of users. Thus, it is difficult for data scientists to obtain realistic data to train their models, even internally within an organization. To address data privacy issues, three main approaches are usually explored [2,3]: 1) non-cryptographic anonymization methods; 2) cryptographic anonymization methods; and, 3) perturbation methods, such as differential privacy. However, 1) leaks private information in most cases; 2) is currently impractical (e.g., using homomorphic encryption) for large data sets; and, 3) degrades data quality making it less suitable for machine learning tasks.

In this paper, we take an orthogonal approach. We generate synthetic data that is realistic enough to replace real data in machine learning tasks. Specifically, we consider multivariate time-series data and, unlike prior work, capture both temporal dependencies and dependencies between attributes. Figure 1 illustrates our approach, called *STAN*. Given real historical data, we first train a CNN-based autoregressive generative neural network that learns the underlying joint data distribution. The trained STAN model can then generate any amount of synthetic data without revealing private information. This realistic synthetic data can replace real data in the training of machine learning models applied to cybersecurity applications, with performance[1] comparable to models trained on real data.

To evaluate the performance of *STAN*, we use both simulated data and a real publicly-available network traffic data set. We compare our method with four selected baselines using several metrics to evaluate the quality of the generated data. We also evaluate the suitability of the generated data as compared to real data in training machine learning models. Self-supervision is commonly used for anomaly detection [11,32]. We consider two such tasks – a classification task and a regression task for detecting cybersecurity anomalies – that are trained on both real and synthetic data.

We show comparable model performance after entirely substituting the real training data with our synthetic data: the F-1 score of the classification task only drops by 2% (99% to 97%), while the mean square error only increases by about 13% for the regression task.

In this paper, we make the following key contributions:

- *STAN* is a new approach to learn joint distributions of multivariate time-series data—data typically used in cybersecurity applications—and then to generate synthetic data from the learned distribution. Unlike prior work, *STAN* learns both temporal and attribute dependencies. It integrates convolutional neural layers (CNN) with mixture density neural layers (MDN) and softmax layers to model both continuous and discrete variables. Our code is publicly available.[2]
- We evaluated *STAN* on both simulated data and a real publicly available network traffic data set, and compared with four baselines.
- We built models for two cybersecurity machine learning tasks using only *STAN* generated data to train and which demonstrate model performance comparable to using real data.

2   Related Work

Synthetic Data Generation. Generating synthetic data to make up for the lack of real data is a common solution. Compared to modeling image data [22], learning distributions on multi-variate time-series data poses a different set of challenges. Multi-variate data is more diverse in the real world, and such data usually has more complex dependencies (temporal and spatial) as well as heterogeneous attribute types (continuous and discrete).

Synthetic data generation models often treat each column as a random variable to model joint multivariate probability distributions. The modeled distribution is then used for sampling. Traditional modeling algorithms [4,13,31] have

---

[1] Here performance refers to a model evaluation metric such as precision, recall, F1-score, mean squared error, etc.

[2] https://github.com/ShengzheXu/stan.git.

the restraint of distribution data types and due to computational issues, the dependability of synthetic data generated by these models is extremely limited. Recently, GAN-based (Generative Adversarial Network-based) approaches augment performance and flexibility to generate data [17,23,34].

However, they are still either restricted to a static dependency without considering the temporal dependence usually prevalent in real world data [28,34], or only partially build temporal dependence inside GAN blocks [14,17]. By 'partially' here, we mean that the temporal dependencies generated by GANs is based on RNN/LSTM decoders, which limits the temporal dependence in a pass or a block that GAN generates at a time. In real *netflow* data, traffic is best modeled as infinite flows which should not be interrupted by blocks of the synthesizer model. Approaches such as [14] (which generates embedding-level data, such as the embedding of URL addresses) and [17] are unable to reconstruct real *netflow* data characteristics such as IP address or port numbers, and other complicated attribute types. We are not aware of any prior work that models both entire temporal and between-attribute dependencies for infinite flows of data, as shown in Table 1.

Furthermore, considering the *netflow* domain speciality, such data should not be simply treated as traditional tabular data. For example, [28] passes the *netflow* data into a pre-trained IP2Vec encoder [27] to obtain a continuous representation of IP address, port number, and protocol for the GAN usage. *STAN* doesn't need a separate training component, and instead successfully learns IP address, port number, protocol, TCP flags characteristics naturally by simply defining the formats of those special attributes.

Table 1. Comparison of STAN with recent similar work

Methods	Captures attribute dependence	Captures temporal dependence	Generates realistic IP addresses/port numbers
WP-GAN [28]	✓		✓
CTGAN [34]	✓		
ODDS [14]	✓	Partially	
DoppelGANger [17]	✓	Partially	
STAN	✓	✓	✓

Autoregressive Generative Models. [21,22] have been successfully applied to signal data, image data, and natural language data. They attempt to iteratively generate data elements: previously generated elements are used as an input condition for generating the subsequent data. Compared to GAN models, autoregressive models emphasize two factors during the distribution estimating: 1) the importance of the time-sequential factor; 2) an explicit and tractable density. In this paper, we apply the autoregressive idea to learn and generate time-series multi-variable data.

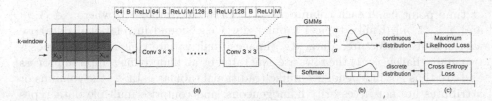

Fig. 2. *STAN* components: (a) window CNN, which crops the context based on a sliding window and extracts features from context; The CNN architecture includes 14 layers where numeric values notes are the number of 3 * 3 convolutional filters; B notes batch normalization layers; ReLU notes activation layers; and M notes max pooling layers. (b) mixture density neural layers and softmax layers learn to predict the distributions of various types of attributes; (c) the loss functions for different kinds of layers.

Mixture Density Networks. Unlike modeling discrete attributes, some continuous numeric attributes are relatively sparse and span a large value range. Mixture Density Networks [6] is a neural network architecture to learn a Gaussian mixture model (GMM) that can predict continuous attribute distributions. This architecture provides the possibility to integrate GMM into a complex neural network architecture.

Machine Learning for Cybersecurity. In the past decades, machine learning has been brought to bear upon multiple tasks in cybersecurity, such as automatically detecting malicious activity and stopping attacks [7,10]. Such machine learning approaches usually require a large amount of training data with specific features. However, training model using real user data leads to privacy exposure and ethics problems. Previous work on anonymizing real data [25] has failed to provide satisfactory privacy protection, or degrades data quality too much for machine learning model training.

This paper takes an different approach that, by learning and generating realistic synthetic data, ensures that real data can be substituted when training machine learning models.

Prior work on generating synthetic network traffic data includes Cao et al. [8] who use a network simulator to generate traffic data while we do not generate data through workload or network simulation; both Riedi et al. and Paxson [24, 26] use advanced time-series methods, however, these are for univariate data, and different attributes are assumed to be independent. We model multivariate data, that is, all attributes at each time step, jointly. Mah [19] models distributions of HTTP packet variables similar to baseline 1 (B1) in our paper. Again, unlike our method, it does not jointly model the network data attributes.

3 Problem Definition

We assume the data to be generated is a multivariate time-series. Specifically, data set $\mathbf{x}$ contains n rows and m columns. Each row $\mathbf{x}_{(i,:)}$ is an observation at time point i and each column $\mathbf{x}_{(:,j)}$ is a random variable j, where $i \in [1..n]$ and $j \in [1..m]$. Unlike typical tabular data, e.g., found in relational database tables, and unstructured data, e.g., images, multivariate time-series data poses two main challenges: 1) the rows are generated by an underlying temporal process and are thus not independent, unlike traditional tabular data; 2) the columns or attributes are not necessarily homogeneous, and comprise multiple data types such as numerical, categorical or continuous, unlike say images.

The data $\mathbf{x}$ follows an unknown, high-dimensional joint distribution $\mathbb{P}(\mathbf{x})$, which is infeasible to estimate directly. The goal is to estimate $\mathbb{P}(\mathbf{x})$ by a generative model $\mathbb{S}$ which retains the dependency structure across rows and columns. Values in a column typically depend on other columns, and temporal dependence of a row can extend to many prior rows. Once model $\mathbb{S}$ is trained, it can be used to generate an arbitrary amount of data, $\mathbf{D}_{synth}$.

Another key challenge is evaluating the quality of the generated data, $\mathbf{D}_{synth}$. Assuming a data set, $\mathbf{D}_{historical}$, is used to train $\mathbb{S}$, and an unseen test data set, $\mathbf{D}_{test}$, is used to evaluate the performance of $\mathbb{S}$, we use two criteria to compare $\mathbf{D}_{synth}$ with $\mathbf{D}_{test}$:

1. Similarity between a metric M evaluated on the two data sets, that is, is $M(\mathbf{D}_{test}) \approx M(\mathbf{D}_{synth})$?
2. Similarity between performance P on training the same machine learning task T, in which the real data, $\mathbf{D}_{test}$, is replaced by the synthetic data, $\mathbf{D}_{synth}$, that is, is $P[T(\mathbf{D}_{test})] \approx P[T(\mathbf{D}_{synth})]$?

4   Proposed Method

We model the joint data distribution, $\mathbb{P}(\mathbf{x})$, using an autoregressive neural network.

The model architecture, shown in Fig. 2, combines CNN layers with a density mixture network [6]. The CNN captures temporal and spatial (between attributes) dependencies, while the density mixture neural layer uses the learned representation to model the joint distribution. More architecture details will be discussed in Sect. 4.2. During the training phase, for each row, STAN takes a data window prior to it as input. Given this context, the neural network learns the conditional distribution for each attribute. Both continuous and discrete attributes can be modeled. While a density mixture neural layer is used for continuous attributes, a softmax layer is used for discrete attributes.

In the synthesis phase, STAN sequentially generates each attribute in each row. Every generated attribute in a row, having been sampled from a conditional distribution over the prior context, serves as the next attribute's context.

4.1   Joint Distribution Factorization

$\mathbb{P}(\mathbf{x})$ denotes the joint probability of data $\mathbf{x}$ composed of n rows and m attributes. We can expand the data as a one-dimensional sequence $\mathbf{x}_1, ..., \mathbf{x}_n$, where each vector $\mathbf{x}_i$ represents one row including the m attributes $x_{i,1}, ..., x_{i,m}$. To estimate the joint distribution $\mathbb{P}(\mathbf{x})$ we express it as the product of conditional distributions over the rows. We start from the joint distribution factorization with no assumptions:

$$\mathbb{P}(\mathbf{x}) = \prod_{i=1}^{n} \mathbb{P}(\mathbf{x}_i | \mathbf{x}_1, ..., \mathbf{x}_{i-1}) \tag{1}$$

Unlike unstructured data such as images, multivariate time-series data usually corresponds to underlying continuous processes in the real world and do not have exact starting and ending points. It is impractical to make a row probability $\mathbb{P}(\mathbf{x}_i)$ depend on all prior rows as in Eq. 1. Thus, a k-sized sliding window is utilized to restrict the context to only the k most recent rows. In other words, a row conditioned on the past k rows is independent of all remaining prior rows, that is, for $i > k$, we assume independence between $\mathbf{x}_i$ and $\mathbf{x}_{<i-k}$. We can thus rewrite the joint distribution $\mathbb{P}(\mathbf{x})$ as the product of the conditional distributions over the prior k rows:

$$\mathbb{P}(\mathbf{x}) = \prod_{i=1}^{k} \mathbb{P}(\mathbf{x}_i | \mathbf{x}_1, ..., \mathbf{x}_{i-1}) \prod_{i=k+1}^{n} \mathbb{P}(\mathbf{x}_i | \mathbf{x}_{i-k}, ..., \mathbf{x}_{i-1}) \tag{2}$$

Note that a suitable value of k needs to be picked based on empirical evidence or domain knowledge. While all the probabilities in the second term on the RHS of Eq. 2 are conditioned on k variables, the same is not true for the probabilities in the first term. To make these consistent, we add zero padding and then symbolically define that $\mathbf{x}_i$ where $i \leq 0$ represents a padding row, as Eq. 3 shows.

$$\mathbb{P}(\mathbf{x}) = \prod_{i=1}^{k} \mathbb{P}(\mathbf{x}_i | \mathbf{x}_{i-k}, ..., \mathbf{x}_1, ..., \mathbf{x}_{i-1}) \prod_{i=k+1}^{n} \mathbb{P}(\mathbf{x}_i | \mathbf{x}_{i-k}, ..., \mathbf{x}_{i-1})$$
$$= \prod_{i=1}^{n} \mathbb{P}(\mathbf{x}_i | \mathbf{x}_{i-k}, ..., \mathbf{x}_{i-1}) \tag{3}$$

The joint distribution of a row can be factorized in two ways: 1) Equation 4 assumes conditional independence of attributes in a row, given all attributes in the previous k rows; 2) Equation 5 makes no conditional independence assumptions of attributes in the same row.

$$\mathbb{P}(\mathbf{x}) = \prod_{i=1}^{n} \prod_{j=1}^{m} \mathbb{P}(\mathbf{x}_{i,j} | \mathbf{x}_{i-k}, ..., \mathbf{x}_{i-1}) \tag{4}$$

$$\mathbb{P}(\mathbf{x}) = \prod_{i=1}^{n} \prod_{j=1}^{m} \mathbb{P}(\mathbf{x}_{i,j} | \mathbf{x}_{i-k}, ..., \mathbf{x}_{i-1}; x_{i,1}, ..., x_{i,j-1}) \tag{5}$$

While (4) provides a good approximation, we found (5) performs slightly better.

During generation, initialization is a key concern for our generative model due to temporal dependence. In order to generate data without supplying any real data or specific seed, we begin the above autoregressive chain process with the marginal distribution $\mathbb{P}(\mathbf{x}_1)$. In practice, as described later in Sect. 4.2, the marginal distribution is approximated by all zero conditional: $\mathbb{P}(\mathbf{x}_1|0)$. Then beginning from the second step, Eq. 4, 5 holds.

4.2   Neural Network Architecture

As shown in Fig. 2, the input window goes through the *convolutional layers* followed by *mixture density neural layers* or *softmax layers* sequentially to learn the joint distribution. We define two *loss* functions for the two distribution modeling layers separately. Algorithms 1 and 2 provide details on model training and data synthesis. Note that the training phase allows for parallelization while the synthesis phase is sequential.

Algorithm 1. Model training process for each attribute j

Input $D_{Historical}$, window size k, attribute type T_j.
Output STAN model $\mathbb{S}_{stan}$;

1: Construct window data
   $X_i^{window} = $ concatenate $X_{i-k},...,X_i$;
   $y_i^{window} = X_i$;
2: **for** epoch in 1 ... EPOCH **do**
3:    $X_i^{window} \mathrel{*}= Mask$
4:    **if** T_j is continuous **then**
5:       $\mathbb{P}_{gmm\_pred} = mdn(wCNN(X_{window}))$;
6:       $loss = nll(\mathbb{P}_{gmm\_pred}, y_{window})$;
7:    **else**
8:       $\mathbb{P}_{softmax\_pred} = softmax(wCNN(X_{window}))$;
9:       $loss = cross\_entropy(\mathbb{P}_{softmax\_pred}, y_{window})$;
10:   **end if**
11:   Using *Adam* optimizer to update $\mathbb{S}_{stan}$ with *loss*;
12: **end for**

Window Convolutional Layers (wCNN). The CNN layers, which we call window CNN since they operate on a sliding window of data, perform a two-dimensional convolution. For one row $\mathbf{x}_i$ the layers capture a rectangular context above the row as shown in Fig. 2 *STAN* uses multiple convolutional layers that preserve the spatial and temporal resolution in a sliding time window box. Each number in Fig. 2 represents the number of $3 * 3$ filters in that layer. Batchnorm, ReLU and max pooling layers are also used, marked as BN, $ReLU$, and M, respectively.

Algorithm 2. Data synthesis process

   Input Trained STAN model $\mathbb{S}_{stan}$.
   Output D_{synth};
1: Init context $X_{\cdot}^{window}$ = marginal sampling()
2: **while** condition(target row number or time stamp) **do**
3:    $X_i^{window} *= Mask$
4:    $P_{pred} = \mathbb{S}_{stan}(X_i^{window})$;
5:    y_{sample} = sample from distribution $\mathbb{P}_{pred}$;
6:    $X_{i+1}^{window} = X_i^{window}[1:,:] + y_{sample}$
7: **end while**

Convolution Mask. Based on which factorization is selected, we have mask A for Eq. 4 and mask B for Eq. 5 (Fig. 3).

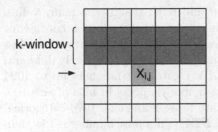 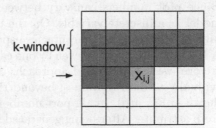

(a) Mask A for conditional independence assumption between attributes in same row

(b) Mask B for no conditional independence assumption in the same row

Fig. 3. Masks for context window convolution

Mixture Density Neural Layer (MDN). Learns a conditional *Gaussian mixture distribution*. It consists of three parallel fully connected layers, modeling $\alpha_i, \sigma_i, \mu_i$ separately, where the parameter α_i represents for the component weights of an *Gaussian mixture model*, and the μ_i and σ_i^2 are the mean and variance parameters of the Gaussian distribution components. The α_i parameters output go to a softmax, so that the weights of all the Gaussian mixture components sum to one.

Loss Functions. We define loss functions for *mixture density neural layer* and *softmax layer* separately. Note that the two losses have different scales, and while multitask learning has its advantages, we match each *mixture density neural component* or *softmax component* with an individual *wCNN component*.

A Negative Log-Likelihood Loss (NLL) is used for the mixture density layers, which predict a group of mixture density parameters that can compose a Gaussian mixture model as Eq. 6: $\alpha_i, \sigma_i, \mu_i$. We use maximum likelihood loss to estimate a true distribution: the label of the input, which is the new row that to be generated, is supposed to have the highest probability in the estimated distribution. Cross entropy loss is used for the softmax layer.

$$NLL(x|\mu, \sigma^2) = -\log \sum \alpha_i * \mathcal{N}(x|\mu_i, \sigma_i^2) \qquad (6)$$

4.3   IP Address and Port Number Modeling

IP addresses and port numbers are key to network traffic data. However, naively modeling them as continuous or discrete variables gives poor results.

$STAN$ specifically learns IP address and port number characteristics.

As shown in Fig. 4(a), $STAN$ treats an IP address as four 256-categorical discrete attributes. With the help of the $STAN$ Mask definition, even though one IP address attribute is split across four intermediate attributes, it is considered together as one variable for attribute dependence.

Since port numbers can vary between 0 and 64K, that is too many values to model as a discrete variable. On the other hand, treating it as a continuous variable would result in inaccuracies especially for well-known ports (those less than 1024), where being off even by one can mean something completely different. Therefore, we take a hybrid approach. $STAN$ treats port numbers up to 1024 individually as discrete values; beyond that it models ports in bins of size 100, as shown in Fig. 4(b). In all, port numbers are represented as a 1670-categorical discrete attribute. After being generated by $STAN$, if a port number is less than 1024 it corresponds to that particular port number, else a port is sampled from a uniform distribution of port numbers in the corresponding bin during post processing.

4.4   Baselines

We selected four different methods to serve as baselines for our method. This range for basic Gaussian Mixture Model, Bayesian Network to two recent deep learning approaches that use GANs for synthetic data generation, which for brevity we refer to as GMM, BN, WPGAN, and CTGAN, respectively. We compare $STAN$ with these baselines and analyze the distribution factorization.

Gaussian Mixture Model (GMM). This assumes all attributes at a particular time step are independent of each other, and further that each row is independent. Thus it can be factorized as following:

$$\mathbb{P}(\mathbf{x}) = \prod_{i=1}^{n} \mathbb{P}(\mathbf{x}_i) \qquad (7a) \qquad\qquad \mathbb{P}(\mathbf{x}_i) = \prod_{j=1}^{m} \mathbb{P}(\mathbf{x}_{i,j}) \qquad (7b)$$

Bayesian Network (BN). As a traditional statistical approach, limited temporal or attributes dependence can be learnt based on the domain knowledge

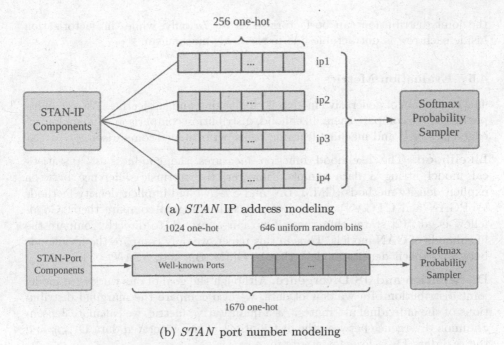

(a) *STAN* IP address modeling

(b) *STAN* port number modeling

Fig. 4. With the benefit of flexible *STAN* continuous and discrete generator architecture, special domain attributes (such as IP address, port, protocol, and TCP flags), can be learned by purely modifying the configure parameters.

from experts. For example, if $\mathbf{x}_{i,j_1}$ is dependent on $\mathbf{x}_{i-1,j_1}$ and $\mathbf{x}_{i,j_2}$, we can write it as a product of the conditional distributions (see Eq. 8). The value $\mathbb{P}(\mathbf{x}_{i,j_1}|\mathbf{x}_{i-1,j_1},\mathbf{x}_{i,j_2})$ is the probability of the j_1 attributes of the i-th observation row, given the $(i-1)$-th j_1 attribute and the i-th j_2 attribute. Considering the edge situation as well as utilizing the Bayes rule, we rewrite the distribution $\mathbb{P}(\mathbf{x}_{i,j_1}|\mathbf{x}_{i-1,j_1},\mathbf{x}_{i,j_2})$ as:

$$\mathbb{P}(\mathbf{x}) = \prod_{i=1}^{n}[\mathbb{P}(x_{i,j_1}|x_{i,j_2},x_{i-1,j_1})\prod_{j=1,j\neq j_1}^{m}\mathbb{P}(x_{i,j})]$$

$$= \mathbb{P}(x_1) \cdot \prod_{i=2}^{n}[\mathbb{P}(x_{i,j_1})\mathbb{P}(x_{i-1,j_1}|x_{i,j_1})\mathbb{P}(x_{i,j_2}|x_{i,j_2})] \qquad (8)$$

$$\cdot \prod_{j=1,j\neq j_1}^{m}\mathbb{P}(x_{i,j})$$

WPGAN [28] utilizes GAN to specifically generate network traffic flow data, while **CTGAN** [34] utilizes GAN to generate general tabular data that contains both discrete and continuous attributes. Both B3 and B4 assume attribute dependence at a certain time step but ignore temporal-wise dependence. Thus the joint distribution can be factorized as Eq. 7a only, while the factorization inside each row is untractable due to the GAN mechanism.

4.5   Evaluation Metrics

The evaluation of generative models is challenging and subjective. We use multiple metrics to compare them: likelihood, distribution comparison, domain knowledge rules test, and machine learning tasks performance comparison.

Likelihood. The likelihood function measures the goodness of a statistical model fitting a data sample. However, the intrinsic difference between explicit density methods (GMM, BN, and $STAN$) and implicit density methods (WPGAN and CTGAN) makes it more challenging to compare them. Goodfellow et al. [12] states that there is no fair approach to directly compare the likelihoods of GAN models. Thus in this paper, we only compare the likelihoods between explicit density models, that is, GMM, BN and, $STAN$.

Distribution and JS Divergence. Although the goal of our work is to model joint distribution of a window of data, we also compare the marginal distributions of the individual attributes. As a quantitative metric, we calculate Jensen-Shannon divergence between the distributions of the generated data $\mathbf{D}_{synth}$ and the real data $\mathbf{D}_{test}$ for each attribute.

Domain Knowledge Test. We use domain knowledge checks to evaluate the synthetic data quality. Since the application data set pertains to network traffic flow, we use several properties that such data needs to satisfy in order to be realistic [28].

In addition to marginal distributions, we also explore network traffic specific distributions such as that of the number of unique destinations (in terms of IP addresses), and of number of bytes per IP address, and compare then with those distributions in real data. Similarly, we compare the top most frequently occurring port numbers.

Machine Learning Application Task. The final goal of generating synthetic data is to build machine learning models without using any real data. To evaluate whether the generated data is able to replace real data in a model training process, we select two tasks used for cybersecurity anomaly detection. One is a classification task while the other is regression; both use self-supervision.

While these tasks may not seem to directly relate to cybersecurity, they are good at detecting anomalies, which may be caused, e.g., by a cybersecurity breach. The basic assumption is that different attributes in network traffic data have underlying dependencies, e.g., between the protocol field and other fields, during normal operation and these relationships may change in the presence of security anomalies. The models are built to capture this normal relationship, and then try to detect any changes in it during deployment.

The first task is predicting the transport protocol field in network traffic flow data, while the second task is to predict the number of bytes in the next network flow. In practice once trained these models are used for marking anomalies when the actual value significantly differs from the real one based on a hyperparameter threshold value. We train a RandomForest model for the classification task, and a neural network model for the regression task.

As the evaluation metric, we use the F1 score, and the mean square error (MSE) for the two tasks, respectively. Since the classification task is an multi-class task, we apply the macro-F1 score which takes the average of all the category F1 scores. This ensures equal treatment of all classes even when the class distribution is skewed, as is likely for transport protocol where TCP dominates. For both tasks we compare the cross-validation performance of the models trained on real and synthetically generated data.

5   Experimental Results

To demonstrate its effectiveness, we train and evaluate $STAN$ on a real network traffic data set. However, to experiment with some architectural variations, we first use a simple simulated data set.

5.1   Understanding STAN Using Simulated Data

We built a simulated dataset with a simple random process whose dependence can be clearly controlled. We simulated a two-variable data distribution with the following formula and sampled 10,000 points data set (X, Y) from it:

$$x_t = 0.9x_{t-1} + 0.1\mathcal{N}$$

and

$$y_t = 0.9x_t + 0.1\mathcal{N},$$

where $\mathcal{N}$ is standard normal distribution. We incorporate both temporal and attribute dependence with two attributes, X and Y.

To train we apply a naive version of $STAN$, that passes the input to mixture density neural layers directly, and GMM on the simulated data set. Then we generated data using both $STAN$-a and $STAN$-b.

To measure how well dependence is captured we use the Pearson correlation coefficient R both for temporal dependence $R(X_t, X_{t-1})$ and attribute dependence $R(X_t, Y_t)$. In addition, we use two machine learning tasks to show that the synthetic data is able to serve as model training source to replace the original data.

Quantitative Evaluation. We evaluated the correlation coefficient R between both temporal dependence $R(X_t, X_{t-1})$ and attribute dependence $R(X_t, Y_t)$. Figure 5 shows the scatter plots of X_t and X_{t-1} from four data sources (simulated data, GMM synthetic data, $STAN$ with mask A synthetic data, and $STAN$ with mask B synthetic data), and Fig. 6 shows that for X_t and Y_t.

Figures 5a and 6a show the dependence in the original (simulated) data, across time and attribute, respectively, that the synthesizer needs to learn. The scatter plots (x_{t-1}, x_t) and (x_t, y_t) show strong linear relationship. Figures 5b and 6b show that independently learned marginal distribution is unable to generate data with temporal and attribute-wise dependence. Figure 5c and 5d show *STAN-a* and *STAN-b* perform similarly and are able to generate data with the same temporal dependence as in the simulated (original) data. However, Fig. 6c and 6d show that *STAN-b*, when explicitly conditioned on the same-row attribute context in its convolution perceptive field, generates better attribute-wise dependence than *STAN-a*.

Since mask A and mask B represent *conditional independence* and *explicit dependence* respectively, we summarize through this observations:

- Both conditional independence and explicit dependence provide reasonable approximation for temporal dependence. The $R(X_t, X_{t-1})$ of simulated data, synthetic data generated using *STAN* mask A, and synthetic data generated using *STAN* mask B are all same (0.9).
- Conditional independence provides a reasonable same-row attribute approximation, while explicit dependence performs better. The $R(X_t, Y_t)$ of simulated data, and synthetic data generated using *STAN* mask B is 0.9; while that of synthetic data generated using *STAN* mask A is 0.7.

Machine Learning Tasks. We also used the simulated data and the corresponding synthetic data for training two machine learning tasks (using the scikit-learn Python library). Then we evaluate the performance (Mean Square Error) of the trained machine learning models on the simulated test data. Simulated training data and simulated test data follow the same distribution.

- **T1:** predict y_t given x_t (row attribute dependence).
- **T2:** predict x_{t+1} given x_t (temporal dependence).

Table 2 shows that a machine learning model trained only on synthetic data generated by *STAN* produces similar test loss as that trained on the original simulated test data.

Table 2. Mean square error of the two tasks

Training data	MSE (T1)	MSE (T2)
Simulated data	0.010	0.01
GMM	0.050	0.05
STAN mask A	0.013	0.01
STAN mask B	0.010	0.01

5.2   Real Network Traffic Data

Data Set. Network traffic data is typically a multivariate time-series. A common format is called *netflow*, where each row represents a unidirectional network

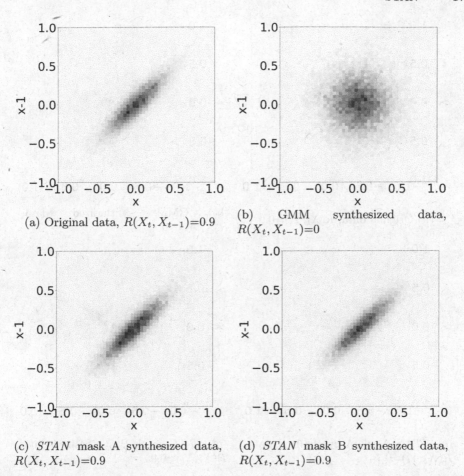

(a) Original data, $R(X_t, X_{t-1})=0.9$

(b) GMM synthesized data, $R(X_t, X_{t-1})=0$

(c) *STAN* mask A synthesized data, $R(X_t, X_{t-1})=0.9$

(d) *STAN* mask B synthesized data, $R(X_t, X_{t-1})=0.9$

Fig. 5. Temporal dependence: (X_t, X_{t-1}) scatter plot of the simulated data and synthetic data with Correlation Coefficients R.

traffic connection or flow. We selected a *netflow* data set for our experiments since it is a good representative format for network traffic data in general. Here we use a large publicly available *netflow* data set. Typically each row consists of the following attributes: timestamp at the end of a flow (te), duration of flow (td), packets exchanged in the flow (pkt), and the corresponding number of bytes (byt), source IP address[3] (sa), destination IP address (da), source port (sp), destination port (dp), flags (flg), and transport protocol (pr). Each row x_i can be expressed as a tuple of $(te_i, byt_i, sa_i, da_i, pr_i$, etc.). Table 3 shows typical attributes, their types and example values.

We apply *STAN* on a publicly available benchmark *netflow* data set, UGR'16 [18], which contains large scale traffic data captured by a Tier-3 ISP cloud service

---

[3] We only consider IPv4 addresses here.

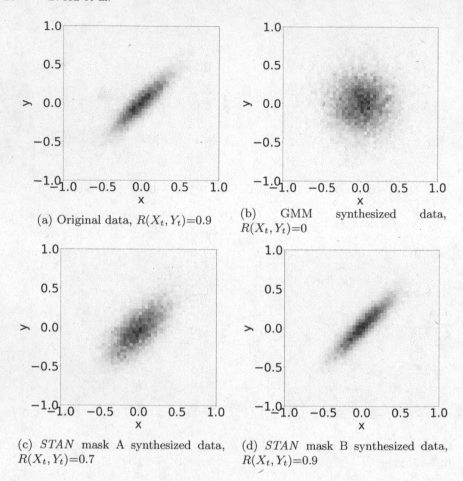

(a) Original data, $R(X_t, Y_t)=0.9$

(b) GMM synthesized data, $R(X_t, Y_t)=0$

(c) *STAN* mask A synthesized data, $R(X_t, Y_t)=0.7$

(d) *STAN* mask B synthesized data, $R(X_t, Y_t)=0.9$

Fig. 6. Attribute dependence: (X_t, Y_t) scatter plot of the simulated data and synthetic data with Correlation Coefficients R.

provider. First, we selected a week of data (April week3) data to focus on. We selected this week this week since it looked interesting in terms of volume of traffic and the number of events marked. Second, we randomly selected flows related to 90 users (essentially IP addresses) based on the number of traffic flows per user distribution. Third, we extracted one day's (Monday) data to be the $\mathbf{D}_{historical}$ and another day (Tuesday) of the same user group and the same week to be the $\mathbf{D}_{test}$. Following this strategy, we selected 1,531,126 samples for the $\mathbf{D}_{historical}$ and 1,952,702 samples for the $\mathbf{D}_{test}$. Ten percent of $\mathbf{D}_{historical}$ are selected out to serve as training validation data $\mathbf{D}_{validation}$.

netflow **Data Processing.** To ensure the trained model is a practical and robust tool to synthesize network traffic flow data, we normalize the raw *netflow* data for ease of processing by the neural network. Also, the neural network predicted values are transformed back into the original scale.

Table 3. Overview of typical attributes in flow-based data.

Attribute	Type	Example
timestamp	continuous	2016-04-11 00:02:15
duration	continuous	0.344
transport protocol	categorical	TCP
source IP address	categorical	85.201.196.53
source port	categorical	19925
dest. IP address	categorical	42.219.145.151
dest. port	categorical	80
bytes	numeric	11238
packets	numeric	11
TCP flags	categorical	.A..SF

The inputs to the neural model are pre-processed to facilitate training. The numerical attributes are min-max scaled; for the categorical attributes, we apply one-hot encoding. Specifically, for the protocol attribute we use a three-way softmax (for TCP, UDP and other). Note that for simplicity we consider only three protocol categories since TCP and UDP are the most prevalent; it would be easy to extend it to more categories if needed. For source and destination port number attributes, we handle well-known and other ports differently as described in Sect. 4.3. Instead of modeling timestamps of individual flows, we model the time deltas between them.

Training Hyperparameters. Our models are trained on four Tesla P100 GPUs using the Pytorch toolbox. From the different parameter update rules tried, the Adam [15] algorithm gives best convergence performance and is used for all experiments. The learning rate schedules were manually set to the highest values that allowed fast convergence: 0.001 for mixture density neural layers and 0.01 for softmax layers. The batch sizes are also manually set for the experiments. For UGR16, we use as large a batch size as that showed quick converge; this corresponds to 512 time windows input per batch. We use pre-processing to prepare data batches that can be trained in parallel and accelerate the training and generation process. For mixture density neural layers, we select 10 as the Gaussian components to be learned based on the cross-validation. For the initial convolution network layer parameters, we sample from a Uniform distribution whose boundary is the standard deviation of the kernel size, i.e. $[-\frac{1}{3*3}, \frac{1}{3*3}]$.

Likelihood. For each data point (each row), we can directly calculate the row likelihood by factorization equations. In our case, explicit density generative models (GMM, BN and $STAN$) clearly define the distribution for each attributes and for those we can evaluate the modeled distribution directly via individual attribute distributions. For simplicity, we discretize continuous variables to validate their negative log-likelihood value for all the baselines and attributes, based on the variable value range and the data set size. In Table 4 we report the negative log likelihood (NLL) of a few attributes as modelled by *STAN* and baselines GMM and BN. Note that we are unable to generate IP addresses and port numbers using GMM and BN, so the NLL for those attributes is not compared in the table. *STAN* produces better results for both continuous and discrete attributes.

Table 4. Attribute negative log likelihood of models evaluated on $\mathbf{D}_{validation}$ (lower is better).

Model	Bytes	Packet	Time duration	Transport protocol
GMM	4.85	3.78	1.81	0.341
BN	3.90	2.62	0.97	0.344
STAN	2.34	1.73	0.59	0.002

Data Synthesis. Once a *STAN* model is trained on $\mathbf{D}_{historical}$, it is able to sequentially generate any length of *netflow* data $\mathbf{D}_{synth}$. As described in Sect. 4, *STAN* starts the generation process by sampling from the trained marginal distribution without any input requirement, and then autoregressively generates row by row conditioned on the prior rows context. To fairly compare to $\mathbf{D}_{test}$, we used *STAN* to generate $\mathbf{D}_{synth}$ that includes 1,208,182 samples for the same set of $\mathbf{D}_{historical}$ users. Note that since we are generating data in one day's range (based on the generated delta time attribute dt and the accumulated timestamp), the total number of samples is not directly determined by any hyperparameter. In the rest of this section, we evaluate the comparability between the $\mathbf{D}_{synth}$ and $\mathbf{D}_{test}$.

Distribution and JS Divergence. Figure 7 shows the individual JS divergence of the marginal distribution of both the continuous and discrete attributes. *STAN* captures the marginal distribution well for most attributes. Even though GMM precisely models the marginal distribution of the training data set, it does not perform as well as *STAN* on the test data set. We believe this is because the marginal distribution over days is non-stationary.

Observation 1: *STAN models the marginal distribution better than baseline GMM.*

Domain Knowledge Test. As described earlier we perform basic sanity checks specified as rules on that need to be satisfied by generated flow-based network data. There are two basic types of rules – those that apply to an individual attribute and those that check relationships between multiple attributes. Note that individual tests on port numbers and protocols are not needed since they are modeled as categorical variables. We highlight five tests here which are summarized in Table 5. *STAN* performs well in all three.

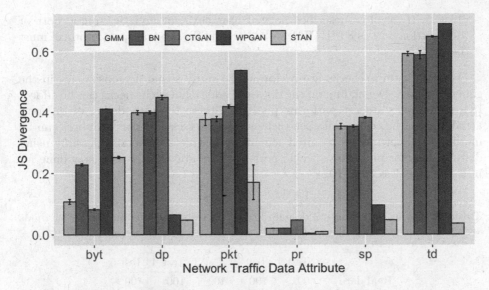

Fig. 7. JS divergence between attribute marginal distribution between $\mathbf{D}_{test}$ and $\mathbf{D}_{synth}$ from *STAN* as well as that from baselines.

- Test 1: Validity of IP address. Source IP address should not be multicast (from 224.0.0.0 to 239.255.255.255) or broadcast (255.xxx.xxx.xxx); Destination IP address should not be of the form 0.xxx.xxx.xxx. (Note that source address can be all zeros, e.g. for DHCP requests.)
- Test 2: Number of bytes/packets minimum size: The minimum value for *pkt* attribute is 1 and the minimum value for *byt* attribute depends on the transport protocol. For a TCP flow packet, the minimum size is 40 bytes (20 bytes for IP header + 20 bytes for TCP header); similarly for a UDP flow packet, the minimum size is 28 bytes (20 bytes for IP header + 8 bytes for UDP header).
- Test 3: Relationship between number of bytes (byt) and number of packets (pkt). Based on the protocol the following relationship exists between these attributes. For a TCP flow,

$$40 * pkt \le byt \le 65535 * pkt$$

Similarly, for a UDP flow,

$$28 * pkt \le byt \le 65535 * pkt$$

The maximum packet size for both TCP (assuming maximum MTU) and UDP is 64K bytes.
- Test 4: If the transport protocol is not TCP, then the flow should not have any TCP flags.

– Test 5: If the flow describes normal user behavior and the source port or destination port is 80 (HTTP) or 443 (HTTPS), the transport protocol must be TCP[4].

Domain specific checks are performed to verify semantic consistency in the generated data. Depending on the data set, additional such checks can be added. If a generated data set performs poorly on these tests, modeling of different attributes in the traffic data set such as IP addresses and port numbers can be modified to capture the required semantic constructs, including possibly using distinct bins for modeling specific port numbers above 1024 as is now done for port numbers less than 1024.

Table 5. Passing percentage of domain knowledge tests. Dash (–) means the methods are not able to generate 'flags' attribute.

	Test 1	Test 2	Test 3	Test 4	Test 5
Real data	99	100	100	100	100
GMM	98	66	48	–	77
BN	98	67	49	–	78
WPGAN [28]	99	75	72	97	99
CTGAN [34]	99	93	74	–	99
STAN	**99**	**93**	**93**	**100**	**99**

Marginal Distributions. We explore marginal distributions of *netflow* attributes in the generated data. Figure 8a shows the distribution of number of unique IP addresses each user communicates with. Typically, such distributions follow a power law distribution [20]. *STAN* generated synthetic data comes closest to the real network data, including the unique IP address maximum occurrence value, the total number of different unique IP addresses, and the distribution curve. Meanwhile, Fig. 8b shows the distribution of the total number of bytes exchanged related to each user (IP address). Again, *STAN* synthetic data follows the real data distribution closer than the baseline methods. It is worth noting that *STAN* learned these distributions without any explicit design choices on our part, e.g., in the loss function. It correctly inferred by marginal distribution by virtue of learning the joint distribution.

We also compare port number distribution between real and synthetic data. Based on the real data, we select the top 5 TCP services and top 3 UDP services, which appear most frequently and the occurrence ratio are greater than 1% in the entire TCP or UDP traffic. Figure 9a shows the occurrence probability of that service in the entire TCP traffic records. We have 80/443 port for HTTP/HTTPS service (Hypertext Transfer Protocol/Hypertext Transfer Protocol Secure), 25

---

[4] There is an effort to move HTTP/S to UDP, referred to as QUIC [1]. We use this test here since there was no QUIC traffic in our netflow dataset. However, this test may not be relevant for a different dataset.

port for SMTP (Simple Mail Transfer Protocol), 53 port for DNS (Domain Name System), 110 port for POP3 (Post Office Protocol, version 3), and 22 port for SSH (Secure Shell). Similarly in Fig. 9b, we have 53 port for DNS service, 161 port for SNMP (Simple Network Management Protocol), and 123 port for NTP (Network Time Protocol). We find *STAN* performs well at generating a port number distribution similar to real data. Further, this implies that *STAN* does a good job of capturing application level traffic, which can be mapped to different ports. While WPGAN with IP2Vec component is the second best, the rest of baselines (GMM, BN, CTGAN) perform poorly.

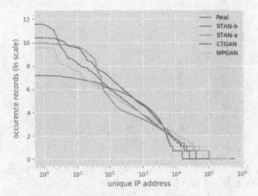

(a) Unique IP address occurrence distribution. The y-axis (ln-scale) is the number of distinct IP addresses contacted.

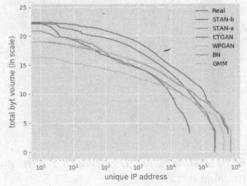

(b) Unique IP volume (*bytes*) distribution. The y-axis (ln-scale) is the total traffic in bytes related to a user (unique IP address).

Fig. 8. IP address characteristics. The x-axis (log-scale) represents unique user (IP address) that occurs in the *netflow* traffic.

Figures 10a and 10b show the marginal distribution of protocol and flags attributes respectively. As expected, TCP is the dominant protocol followed by

UDP. For simplicity, we group the other protocols as 'other'. For both protocol and flags, *STAN* generated data shows a similar distribution to real data.

Observation 2: *Compared to baselines, STAN can learn the IP and port characteristics better without using domain knowledge or other design tuning.*

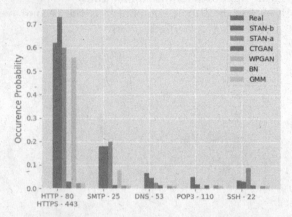

(a) Top 5 most frequently occurring TCP ports and their related service

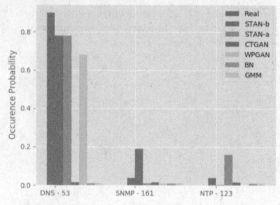

(b) Top 3 most frequently occurring UDP ports and their related service

Fig. 9. Port number characteristics.

Cybersecurity Application Tasks. Finally, we test our synthetic data on two cybersecurity machine learning applications, to detect anomalies using self-supervision. One of the tasks is a classification problem, and the other is a regression problem. The goal is to figure out whether it is possible to fully substitute real data with synthetic data for training machine learning models.

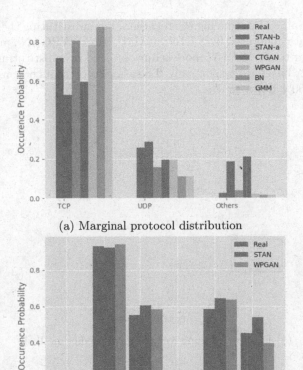

(a) Marginal protocol distribution

(b) TCP flag distribution, includings six flags:
URG, ACK, PSH, RES, SYN, and FIN

Fig. 10. Transport protocol and TCP flags characteristics.

A series of models are trained on real test data. We start our training from using a complete D_{test} (real data) and successively decrease the amount of real data until no data from D_{test} is used. Another series of models are trained similarly using the real test data; however, instead of simply removing certain amount of data from D_{test}, we substitute the indicated amount of data with our synthetic data D_{synth}, so that the total amount of data is kept unchanged.

In the following two tasks, we use D_{test}, which is unseen and never used in the synthesizer training process. For the synthetic data D_{synth}, every synthesizer model generates five sets of synthetic data sample, so we can compute error bars. Five-fold cross validation is used to get a robust estimate of the measurements.

Task1: Protocol Forecasting. Figure 11a shows the F-1 scores achieved by Random Forest models. There are six sets of models. 'Real-Data': these are random forest models trained by reducing the real data; '*STAN*': these are random forest models trained by reducing the real data, but substituting the reduced data by synthetic data generated by *STAN*; 'GMM', 'BN', 'WPGAN', and 'CTGAN': these are similar to the '*STAN*' models but obtained by substituting the reduced data by the four baselines respectively. The x-axis represents how much real data is used from 100% down to 0%.

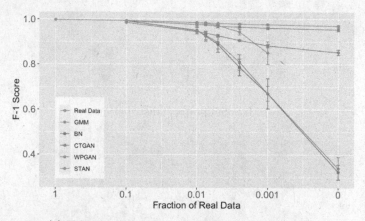

(a) F1-score of same-row Protocol prediction Task

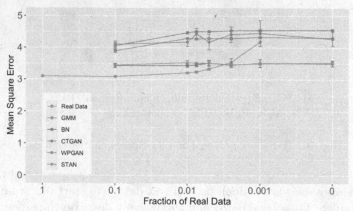

(b) Mean Square Error of *bytes* Value Forecasting Task

Fig. 11. Real application task performance.

If we only use real data, the F1 score drops from 0.99 down to 0.97 as the amount of data decreases. Clearly, with no real data, we are unable to train a model. When we substitute real data with that generated by the baselines, the performance drops even quicker, because they do a poor job of capturing the temporal and attribute dependence. Even in the absence of any real data, data generated by *STAN* results in an F1 score of 0.97, where the drop in performance is only 2%. That is, the model built with only synthetic data retains 98% of the performance of the all real data trained model.

Task2: *bytes* **Value Forecasting.** follows a similar setup of experiments as Task1. Figure 11b shows the mean square error achieved by a neural network regression model. The plot shows that *STAN* and Bayesian network (BN) outperform the other three baseline models. Building a Bayesian network with domain knowledge typically performs better than GANs [34].

In our experiments, BN is optimized specifically for the *bytes* sequential value. However, *STAN* has two advantages over the Bayesian network. First, users do not need the domain knowledge required for Bayesian network implementation. Secondly, there is no inherent bias attributable to an expert unlike traditional Bayesian networks. Similar to the first task, the penalty for using only *STAN* generated data (with no real data) is low, an increase of 13% in the mean square error.

Observation 3: *Compared to BN, STAN performs better on task1 and as well on task2 without requiring any domain knowledge.*

Observation 4: *Even with 0% real data, STAN models task1 and task2 with only a small drop in accuracy.*

6   Conclusion and Future Work

This paper presents the design and implementation of *STAN*, a novel, flexible and robust approach to learn the distribution of complex multivariate time-series data distributions. Compared to existing approaches, *STAN* is novel in several aspects. First, *STAN* learns the joint distribution over both temporal dependency and attribute dependency. Second, *STAN* is able to generate data with any combination of continuous and discrete attributes. Furthermore, our architecture specifically supports generation of IP addresses and port numbers, which makes it particularly suitable for network traffic data. We perform a thorough evaluation of *STAN* comparing it with four baselines using several performance measures as well as on two cybersecurity machine learning tasks.

In the future, we plan to conduct a general and robust privacy preserving evaluation of the generated data. In particular, we plan to empirically validate privacy of training data, that is, no training data is leaked in the generated synthetic data by conducting privacy attacks, such as membership inference attacks [30], and also ensuring there is no training data memorization in our model [9]. Other future work includes: (1) Experimenting with larger filters to validate modeling of longer term temporal dependency in training data. (2) Generate anomalous (attack) data in addition to normal data. (3) explore the best updating rate for re-learning the data synthesizer on historical data $D_{historical}$ on an ongoing basis; (4) conduct more semantic or statistic checking with regards to the fungibility of synthetic data with real data; and (5) support training with and generation of IPv6 addresses.

Acknowledgment. This work is supported in part by the Commonwealth Cyber Initiative (CCI) and US NSF grant DGE-1545362. Any opinions, findings, and conclusions or recommendations expressed in this material are those of the author(s) and do not necessarily reflect the views of the sponsors.

References

1. Quic. https://en.wikipedia.org/wiki/QUIC. Accessed 20 Nov 2020
2. Aggarwal, C.C., Yu, P.S.: A general survey of privacy-preserving data mining models and algorithms. In: Aggarwal, C.C., Yu, P.S. (eds.) Privacy-Preserving Data Mining. Advances in Database Systems, vol. 34, pp. 11–52. Springer, Boston (2008). https://doi.org/10.1007/978-0-387-70992-5_2
3. Al-Rubaie, M., Chang, J.M.: Privacy-preserving machine learning: threats and solutions. IEEE Secur. Privacy **17**(2), 49–58 (2019)
4. Aviñó, L., Ruffini, M., Gavaldà, R.: Generating synthetic but plausible healthcare record datasets. arXiv preprint arXiv:1807.01514 (2018)
5. Berman, D.S., Buczak, A.L., Chavis, J.S., Corbett, C.L.: A survey of deep learning methods for cyber security. Information **10**(4), 122 (2019)
6. Bishop, C.M.: Mixture density networks (1994)
7. Buczak, A.L., Guven, E.: A survey of data mining and machine learning methods for cyber security intrusion detection. IEEE Commun. Surv. Tutor. **18**(2), 1153–1176 (2015)
8. Cao, J., Cleveland, W.S., Gao, Y., Jeffay, K., Smith, F.D., Weigle, M.: Stochastic models for generating synthetic HTTP source traffic. In: IEEE INFOCOM 2004, vol. 3, pp. 1546–1557. IEEE (2004)
9. Carlini, N., Liu, C., Erlingsson, Ú., Kos, J., Song, D.: The secret sharer: evaluating and testing unintended memorization in neural networks. In: 28th USENIX Security Symposium (USENIX Security 2019), pp. 267–284 (2019)
10. Catania, C.A., Garino, C.G.: Automatic network intrusion detection: current techniques and open issues. Comput. Electr. Eng. **38**(5), 1062–1072 (2012)
11. Chen, X., Li, B., Shamsabardeh, M., Proietti, R., Zhu, Z., Yoo, S.: On real-time and self-taught anomaly detection in optical networks using hybrid unsupervised/supervised learning. In: 2018 European Conference on Optical Communication (ECOC), pp. 1–3. IEEE (2018)
12. Goodfellow, I., et al.: Generative adversarial nets. In: Advances in Neural Information Processing Systems, pp. 2672–2680. MIT Press, Cambridge (2014)
13. Graham, C.: Differentially private spatial decompositions. In: 2012 IEEE 28th International Conference on Data Engineering (ICDE) (2012)
14. Jan, S.T., et al.: Throwing darts in the dark? Detecting bots with limited data using neural data augmentation (2020)
15. Kingma, D.P., Ba, J.: Adam: a method for stochastic optimization. arXiv preprint arXiv:1412.6980 (2014)
16. Kwon, D., Kim, H., Kim, J., Suh, S.C., Kim, I., Kim, K.J.: A survey of deep learning-based network anomaly detection. Cluster Comput. **22**(1), 949–961 (2017). https://doi.org/10.1007/s10586-017-1117-8
17. Lin, Z., Jain, A., Wang, C., Fanti, G., Sekar, V.: Using GANs for sharing networked time series data: challenges, initial promise, and open questions. In: Proceedings of the ACM Internet Measurement Conference, pp. 464–483 (2020)

18. Maciá-Fernández, G., Camacho, J., Magán-Carrión, R., García-Teodoro, P., Therón, R.: UGR'16: a new dataset for the evaluation of cyclostationarity-based network IDSs. Comput. Secur. **73**, 411–424 (2018)

19. Mah, B.A.: An empirical model of http network traffic. In: Proceedings of INFO-COM 1997, vol. 2, pp. 592–600. IEEE (1997)

20. Metcalf, L., Casey, W.: Chapter 3 - Probability models. In: Metcalf, L., Casey, W. (eds.) Cybersecurity and Applied Mathematics, pp. 23–42. Syngress, Boston (2016). https://doi.org/10.1016/B978-0-12-804452-0.00003-8. http://www.sciencedirect.com/science/article/pii/B9780128044520000038

21. Van den Oord, A., Kalchbrenner, N., Espeholt, L., Vinyals, O., Graves, A., et al.: Conditional image generation with PixelCNN decoders. In: Advances in Neural Information Processing Systems, pp. 4790–4798 (2016)

22. van den Oord, A., Kalchbrenner, N., Kavukcuoglu, K.: Pixel recurrent neural networks. arXiv preprint arXiv:1601.06759 (2016)

23. Park, N., Mohammadi, M., Gorde, K., Jajodia, S., Park, H., Kim, Y.: Data synthesis based on generative adversarial networks. Proc. VLDB Endow. **11**(10), 1071–1083 (2018)

24. Paxson, V.: Fast, approximate synthesis of fractional gaussian noise for generating self-similar network traffic. ACM SIGCOMM Comput. Commun. Rev. **27**(5), 5–18 (1997)

25. Razak, S., Hafizah, N., Al-Dhaqm, A.: Data anonymization using pseudonym system to preserve data privacy. IEEE Access **8**, 43256–43264 (2020)

26. Riedi, R.H., Crouse, M.S., Ribeiro, V.J., Baraniuk, R.G.: A multifractal wavelet model with application to network traffic. IEEE Trans. Inf. Theory **45**(3), 992–1018 (1999)

27. Ring, M., Dallmann, A., Landes, D., Hotho, A.: Ip2vec: learning similarities between IP addresses. In: 2017 IEEE International Conference on Data Mining Workshops (ICDMW), pp. 657–666. IEEE (2017)

28. Ring, M., Schlör, D., Landes, D., Hotho, A.: Flow-based network traffic generation using generative adversarial networks. Comput. Secur. **82**, 156–172 (2019)

29. RiskIQ Inc.: The evil internet minute 2019 (2019). https://www.riskiq.com/infographic/evil-internet-minute-2019

30. Shokri, R., Stronati, M., Song, C., Shmatikov, V.: Membership inference attacks against machine learning models. In: 2017 IEEE Symposium on Security and Privacy (SP), pp. 3–18. IEEE (2017)

31. Sun, Y., Cuesta-Infante, A., Veeramachaneni, K.: Learning vine copula models for synthetic data generation. In: Proceedings of the AAAI Conference on Artificial Intelligence, vol. 33, pp. 5049–5057 (2019)

32. Vyas, A., Jammalamadaka, N., Zhu, X., Das, D., Kaul, B., Willke, T.L.: Out-of-distribution detection using an ensemble of self supervised leave-out classifiers. In: Ferrari, V., Hebert, M., Sminchisescu, C., Weiss, Y. (eds.) ECCV 2018. LNCS, vol. 11212, pp. 560–574. Springer, Cham (2018). https://doi.org/10.1007/978-3-030-01237-3_34

33. WhiteHouse: The cost of malicious cyber activity to the U.S. economy (2018). https://www.whitehouse.gov/wp-content/uploads/2018/03/The-Cost-of-Malicious-Cyber-Activity-to-the-U.S.-Economy.pdf

34. Xu, L., Skoularidou, M., Cuesta-Infante, A., Veeramachaneni, K.: Modeling tabular data using conditional GAN. In: Advances in Neural Information Processing Systems, pp. 7333–7343 (2019)

Machine Learning for Fraud Detection in E-Commerce: A Research Agenda

Niek Tax[(✉)], Kees Jan de Vries, Mathijs de Jong, Nikoleta Dosoula,
Bram van den Akker, Jon Smith, Olivier Thuong, and Lucas Bernardi

Booking.com, Amsterdam, The Netherlands
{niek.tax,kees.devries,mathijs.dejong,nikoleta.dosoula,bram.vandenakker,
jon.smith,olivier.thuong,lucas.bernardi}@booking.com

Abstract. Fraud *detection* and *prevention* play an important part in ensuring the sustained operation of any e-commerce business. *Machine learning* (ML) often plays an important role in these anti-fraud operations, but the organizational context in which these ML models operate cannot be ignored. In this paper, we take an organization-centric view on the topic of fraud detection by formulating an *operational model* of the anti-fraud departments in e-commerce organizations. We derive 6 research topics and 12 practical challenges for fraud detection from this operational model. We summarize the state of the literature for each research topic, discuss potential solutions to the practical challenges, and identify 22 open research challenges.

Keywords: Fraud detection · Anti-fraud operations · Research agenda

1 Introduction

E-commerce is an important and rapidly growing sector that has tripled its share of the world GDP from 0.5% to >1.5% in the past decade [55]. This surge in economic importance is accompanied by a rapid increase in the total cost of global cybercrime, which increased from $445 billion in 2014 to >$600 billion in 2017 [52]. Fraud and cybercrime in the e-commerce domain spans a variety of fraud types, such as *fake accounts* [16], *payment fraud, account takeovers* [39], and *fake reviews*.

Machine learning (ML) plays an important role in the detection, prevention, and mitigation of fraud in e-commerce organizations. Publicly-known examples include Microsoft [56], LinkedIn [83], and eBay [63]. In practice, fraud detection ML models in e-commerce organizations do not operate in isolation, but they are embedded in a larger *anti-fraud department* that also employs *fraud analysts* or *fraud investigators* who perform case investigations and proactively search for fraud trends. This requires fraud detection models to be embedded in the way of working and daily operations of an anti-fraud department.

While the existing literature on fraud detection is extensive, to the best of our knowledge there is currently no work that provides an explicit formulation

G. Wang et al. (Eds.): MLHat 2021, CCIS 1482, pp. 30–54, 2021.
https://doi.org/10.1007/978-3-030-87839-9_2

of the daily operations of anti-fraud departments. This creates a gap between academic work on fraud detection and practical applications of fraud detection in industry. Furthermore, this makes it more difficult to assess whether novel fraud detection methods fit into the practical way-of-working in fraud departments, or whether they address practically relevant challenges.

In this paper, we describe the operational model of an anti-fraud department. We use this operational model to derive a set of practically relevant research topics for fraud detection. For each research topic, we summarize the state of the literature and put forth a set of *open research challenges* that are formulated from a practical angle. The main aim of this paper is to put forward a research agenda of open challenges in fraud detection.

This paper is structured as follows. In Sect. 2 we introduce and discuss the operational model of fraud detection from an organizational point of view, discuss the role that machine learning plays in it, and derive research topics from it. In Sect. 3 to Sect. 8, we zoom in on each of the individual research topics that we introduce in Sect. 2. In each of those sections, we discuss one research topic, list *practical considerations from industry experience*, *summarize the current state of the literature*, and *formulate open research challenges*. We conclude this paper in Sect. 9.

2 An Operational Model of an Anti-fraud Department

In this section, we introduce our operational model (Fig. 1) of the daily operations of anti-fraud departments in e-commerce organizations. We highlight the role of machine learning in the daily operations of anti-fraud departments and derive research topics and practical challenges.

2.1 Entities and Relations in the Operational Model

E-Commerce Platform. Online service where *users* can buy and sell products (e.g., Amazon, Booking.com, or Zalando).

Users. *Genuine users* perform legitimate transactions (e.g., purchases or sales) on the *e-commerce platform*. *Fraudulent users* are wrongful or criminal actors who intend to achieve financial or personal gains through fraudulent activity on the *e-commerce platform*. Examples of such fraudulent activity: purchase attempts with stolen credit cards, abuse of marketing initiatives (e.g., incentive programs), registering fake accounts (e.g. merchant accounts or user accounts), phishing, or other attempts at account take-overs. Users interact with the *e-commerce platform* ①, which in turn, generates *data* ②.

Data. Is generated by the *e-commerce platform* as a result of user interactions. From the ML viewpoint, data can be transformed into *features* and *labels*. Features represent relevant behavior (e.g., browsing, purchasing, messaging, or managing accounts), or business entities (e.g., purchases, products, or users) of fraudulent and legitimate users. Labels indicate whether or not

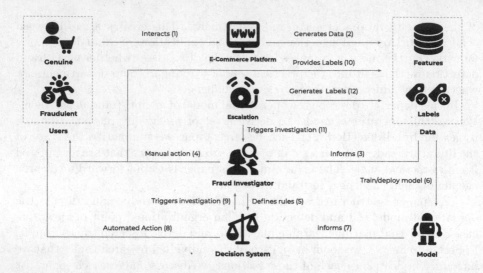

Fig. 1. A model of the daily operations of an anti-fraud department in an e-commerce organization.

behavior or an entity is fraudulent. Labels often result from investigation results of *fraud investigators* ⑩. Sometimes, labels arrive through *external escalations* ⑫, e.g., through notifications of fraud from credit card issuers.

Fraud Investigator. Professionals who investigate suspected fraud cases, using the *data* ③. These suspected fraud cases might originate from internal escalations ⑪ (e.g., complaints through customer service), or the *decision system* triggered an investigation ⑨. For fraud that they find they take *remediating actions* ④ (e.g., canceling orders or blocking users) and/or *preventative steps* by defining *rules* ⑤ that are aimed to identify similar fraud in the future, which are used in the *decision system*.

Decision System. A system that can take concrete actions for instances. *Instances* arise from specific *user requests*, e.g., the purchase of a product, or registration of an account. Instances require a decision on what action to take, e.g., *no intervention, to request additional verification*, or to fully *block* the user's request. The decision system can take automatic action ⑧, or trigger an investigation by a *fraud investigator* ⑨ (who could then take manual action through ④). Actions can either be *synchronous* (i.e., blocking the user request) or *asynchronous* (i.e., without blocking the user request). The decision system decides on its actions by combining *ML models* ⑦ and *rules* ⑤. In addition, some use cases require *exploration*, e.g., by occasionally triggering investigations on instances where there is high uncertainty on whether they are fraudulent.

Model. A machine learning model that aims to distinguish between fraudulent and genuine users. This model is trained ⑥ on the *data*.

Table 1. An overview of the *research topics* that we derived from Fig. 1, the *practical challenges*, and *solution areas* in the literature that relate to them.

Research topic	Connection to Fig. 1	Practical challenges	Solutions areas
Investigation support	③, ⑨, ⑪	Limited capacity	Explainable AI
		Outcome accuracy	Multiple instance learning
		Grouped investigations	Network learning
Decision-making	⑤, ⑦, ⑧, ⑨	Risk management	Probability calibration
		Consequential actions	Cost-sensitive learning
		Combining rules & ML	AI fairness
			Rule-based systems
			Uncertainty quantificat7on
Labels	⑧, ⑩, ⑫	Selection bias	Learning under selection bias
			Multi-armed bandits
Concept drift	⇒(①, ②, ⑥, ⑦, ⑧)	Adversarial drift	Concept drift adaptation
	⇒(①, ②, ⑥, ⑦, ⑨, ④)	Natural drift	Adversarial robustness
		Upstream models	Anomaly detection
ML model–investigator interaction	⇒(⑥, ⑦, ⑨, ⑩)	Explore/exploit trade-off	Active learning
			Guided learning
			Weak supervision
Model	⑥	Model reliability	Model verification
			Automated data validation
			Deployment best practices

2.2 Research Topics

We now discuss the *research topics* that arise from Fig. 1. These *research topics* form the basis of the remainder of this paper, where we dedicate one section per research topic, list their concrete *practical challenges*, provide a summary of existing *solution areas in the literature*, and identify *open research challenges*.

Table 1 summarizes all research topics, their connection to Fig. 1, their practical challenges, and the solution areas in the literature. These connections are either a *set of edges*, or a *path of edges* in Fig. 1. In the latter case, ⇒(Ⓧ, Ⓨ) denotes a path consisting of edges X and Y. Below we introduce the research topics and highlight the practical challenges in **bold**.

Investigation support. Fraud investigations (triggered by ⑨ or ⑪) are performed based on evidence from the data ③. *Fraud investigators* have **limited time capacity**, and to avoid alarm fatigue, ⑨ and ⑪ must yield a high precision. Furthermore, investigators must be enabled and supported to reach decisions *efficiently* and **accurately**. There are several opportunities for machine learning to play a role in supporting these investigations and in the evidence-gathering process that they entail. For example, some fraud cases are highly similar because they are part of the same attack (e.g., they might be performed from the same IP address). Ideally, these are **grouped into a single investigation**, to minimize the number of investigations, and to provide context to the *fraud investigator* during the investigation. We summarize and discuss this research topic in Sect. 3.

Decision-making. Relations ⑤, ⑦, ⑧, and ⑨ show the role of the *decision system*, which is tasked to decide *which* instances to take action on by **combining the output of the ML model and rules** that were created by the *fraud investigators*. The *decision system* is also tasked to decide *how* to take action: either *automatically and immediately* or by sending the case for further review to *fraud investigators*. These decisions should be made with the aim to **manage risk**, i.e., possible risks of negatively impacting genuine users should be traded-off with the possible risk of failing to block fraud. The preventative or remediating actions (e.g., disabling accounts, or stopping purchases) have **great consequences** to the user by design. Therefore, it is essential to limit false positives and to take fairness into account. In Sect. 4, we discuss how the research areas of *cost-sensitive learning*, *AI fairness*, and *uncertainty quantification* offer partial solutions to these challenges and formulate remaining open challenges.

Labels. The two sources of labels are *fraud investigators* ⑩, and automatic escalations ⑫. These mechanisms introduce **selection bias** through delay or incompleteness of labels. In addition, automated actions ⑧ that block suspected fraud mask labels from automatic escalations. We discuss in Sect. 5 how *learning under selection bias* and *multi-armed bandits* offer solutions and we formulate open challenges.

Concept drift. The cycles ⇒(①, ②, ⑥, ⑦, ⑧) and ⇒(①, ②, ⑥, ⑦, ⑨, ④) show how the actions of the *decision system* or the *fraud investigator* impact *fraudulent users*. The fraudulent user may consequently adapt their behavior, i.e., **adversarial drift**. However, the behavior of *genuine users* can also change, i.e., **natural drift**. Moreover, several decisions may be taken at different stages in the life-cycle of a business entity, resulting in **upstream models**. We discuss methods to deal with the adaptivity that this requires from the *ML model* in Sect. 6.

ML-investigator interaction. The cycle ⇒(⑥, ⑦, ⑨, ⑩) highlights the ability of fraud analysts to provide labels to aid ML models. One objective is to investigate the most suspicious instances (i.e., *exploitation*). A contrasting objective is to investigate those instances that are expected to be the most informative to the model (i.e., *exploration*). This creates an **explore/exploit trade-off** regarding which instances are presented to the investigator through ⑨. We discuss the aspects involved in the interaction between the ML model and *fraud investigator* in Sect. 7.

Model. Relation ⑥ concerns the training and deployment of the *ML model*. The fraud detection setting has particular requirements for model **deployment** and **monitoring**, which we discuss in Sect. 8.

3   Investigation Support

Relations ③, ⑨, and ⑪ in Fig. 1 describe the investigations of potential fraud instances that were either found proactively, presented by the *decision system*, or escalated. *Fraud investigations* are often time-consuming and require a high amount of experience and expertise. Much of the time of a fraud investigation goes to gathering evidence and documenting the decision with relevant evidence.

The number of investigations that can be processed is limited because fraud investigations are time-consuming. Therefore, any support from machine learning in aiding the evidence gathering and decision support is of great benefit. The aim is to make evidence gathering *more efficient* and *more effective*, respectively resulting in the ability to *process more investigations* and to *increase the accuracy of the investigation outcome*.

3.1 Summary of the Literature

Explainable AI methods and visualizations thereof [3] can provide decision support to the fraud investigator when embedded in the user interface of investigation tools. Weerts et al. [84] found no strong evidence that SHAP model explanations increase the *accuracy* and *efficiency* of fraud investigators' decision-making. However, in many fraud detection systems, feature interactions are important to fraud detection accuracy. Therefore, one can hypothesize that rule-based explanations such as *anchors* [66], which in contrast to SHAP explain the predictions in terms of rules over multiple features rather than in contribution-scores of individual features, might be better suited for the fraud detection setting. More generally, more research is needed into how fraud investigators can be best supported through model explanations.

In the limited research on the use of model explanations for decision support of the fraud investigator, some work (e.g., [8]) uses crowdsourced labelers (e.g., from Amazon Mechanical Turk) for the experiment. The *fraud investigators* in industry are highly trained, therefore, it is questionable whether empirical results that are obtained with untrained crowdsourced participants transfer to a real-life setting with highly-trained fraud professionals.

In contrast to *local interpretability* methods that explain individual predictions, *global interpretability* methods provide insight into how a model as a whole makes its decisions and might be useful to increase the overall trust of fraud investigators and other stakeholders from the anti-fraud department in the ML model. To the best of our knowledge, empirical work on whether global interpretability methods increase model trust in fraud detection settings is lacking.

Multiple instance learning (MIL) allows ML models to classify whole groups of instances at once instead of single instances. In many fraud use cases, the fraudulent user targets multiple business entities through repeated actions on the e-commerce platform. For example, the same fraudster might perform multiple attempts to compromise accounts. In such cases, performing *group-level investigations*, i.e., investigating multiple instances related to that same actor, yields more appropriate evidence and leads to taking action on more instances per fixed unit of time. MIL is the ML-counterpart of group-level investigations, where individual instances are grouped in Bags, e.g., multiple energy states (instance) of a molecule (bag) in drug discovery, or multiple segments (instance) of an image (bag). Surveys of MIL can be found in [4, 13].

Multiple instance learning methods can aid *group-level investigations* of fraud investigators by identifying groups of instances that could be fraudulent. The bags that are presented to fraud investigators must be relevant, i.e., they must contain a sufficient share of fraud. This is addressed by appropriately defining bag-level labels [13, 24]. There is often a trade-off between bag-level and instance-level model performance [13]. In manual investigations, the former might be more important, while the latter might be more important for automated decisions.

Network learning is closely related to grouped investigations. Graph-based visualizations can show the fraud investigator which instances are connected based on some identifier, e.g., based on *IP address* or *e-mail address*. Like group-level investigations, the graphs provide the fraud investigator with visual information on which instances are connected to some identifier, and which might therefore possibly also be fraudulent. Network learning [46] and graph neural networks [31] are ML counterparts of graph-based investigations. Such models can aid the fraud investigator by identifying graph nodes or subgraphs where the fraud investigator is likely to find fraud.

3.2 Open Research Challenges

Challenge 1: Model explanations for decision support. There is limited research on the effect of model explanations on the *fraud investigator*'s decisions quality and efficiency. It is unclear if these explanations sometimes can bias decisions. It is also unclear what types of model explanations empirically would be most helpful to the fraud investigator, and whether or not this depends on aspects like the application domain, or the experience level of the fraud investigator.

Challenge 2: Multiple instance learning. As pointed out in [13], most of the current literature on MIL covers applications in biology and chemistry, computer vision, document classification, and web mining. To our knowledge, the area of fraud detection in e-commerce, especially the interaction with fraud investigators, has received little attention in the literature, although related topics like the detection of fraudulent financial statements have been discussed in [41], as well as HTTP network traffic in [59]. A particularly interesting challenge is how to compose the bags. In practice, this is often done using characteristics of the user actions (e.g., the IP address and date of the user action cf. [85]). We are not aware of MIL literature that addresses bag construction, especially in the context of fraud detection in e-commerce.

4 Decision-Making

Relations ⑧ and ⑨ in Fig. 1 show that the *decision system* can take *automated action* or *trigger an investigation* for the instances that it suspects of fraud. The *consequences* of wrong decisions can be severe. False positives respectively lead to adding friction for or blocking a genuine user, or wasting the time of fraud investigators. False negatives result in allowing fraudulent behavior. Another, less severe, wrong decision is to trigger an investigation for a fraud case instead of automatically blocking it. It is the decision system's task to *manage risk* by appropriately trading off risks of different types of wrong decisions, by *combining ML and heuristic rules* that are developed by fraud investigators.

Relations ⑤, ⑦ in Fig. 1 highlight the two types of information sources that the *decision system* has available to make its decisions: *ML models* and *rules*. Models and rules aim to complement each other, and it is the task of the *decision system* to aggregate them into a single action when models and rules might recommend different actions for the same instance.

4.1 Summary of the Literature

Probability calibration aims to transform the model output of a classifier in such a way that the predicted model score approximately matches the probability of an instance belonging to the positive class. Calibrated model scores play an important role in *managing risk* in decision-making, as it is the prerequisite of some *cost-sensitive learning* techniques (see below). They also enable the calculation of *expected values* of *key performance indicators* of the business.

Methods for probability calibration include Platt scaling [61], Beta calibration [42], and isotonic regression [87]. Such methods require a *calibration set* of data that is held out from the training data. The adversarial concept drift (see Sect. 6) in fraud settings and the sometimes rapidly changing prevalence (i.e., fraud rate) bring difficulties in obtaining model scores that are close to probabilities *in the latest production data* and not just on the calibration set.

Cost-sensitive learning takes the misclassification costs into account and thereby enables making decisions that minimize the expected cost of fraud to the business operations, rather than simply minimizing the number of classification errors [47]. This addresses the challenge of *managing risk* in decision-making. The application of cost-sensitive learning requires the estimation of the *costs* (or benefit) to the business of the four cells of the confusion matrix: the *false positive, false negative, true positive*, and *true negative*. The cost of a *false positive* could for example be the missed income from a blocked transaction, while the cost of a *false negative* could be the financial cost of that fraud instance.

Practical difficulties often arise because some aspects of these costs can be *difficult to quantify or measure*, such as the *reputational damage* to the business in the case of misclassification. More research is needed on guidelines and frameworks for *how* to design cost functions for cost-sensitive learning when some aspects of the costs are not easily quantified financially.

Once cost functions are in place, the expected costs can be calculated trivially if the classifier returns calibrated scores; alternatively, a method like *empirical thresholding* [77] can be used to minimize the expected costs under uncalibrated model scores. The label might not be available in cases where a fraud attempt was blocked by an automated action. In that case, one might make use of the control group or some other source of unbiased data (see Sect. 5) to optimize the threshold with regard to the cost function.

AI fairness is an important topic in fraud detection because misclassification can have *severe consequences* – a false positive often harms genuine users, like purchases being canceled or accounts being disabled. Since any decision taken in fraud detection potentially directly effects a real person, attention should be paid to ensure fairness and mitigate fairness issues where they may exist.

The body of literature covering fairness in ML is extensive [53], including considerations for its practical applications in production systems [9,34]. While not much fraud-specific applications research has been done in ML fairness, the problem can be cast more generally as a supervised classification setting where positive predictions signify actions taken against individuals. Similar aspects can be found in mortgage default prediction [32] or recidivism prediction [17]. One particular challenge to e-commerce organizations looking to achieve ML fairness is *low observability* on certain protected attributes – dimensions such as race or sex are often not explicitly collected in e-commerce platforms, rendering some attribute-dependent ML fairness methods [82] inapplicable.

The presence of an adversary (fraudster) in the fraud context presents a unique challenge to the implementation of fairness controls. Protected attributes can be spoofed by dishonest actors, for example, by using a VPN to pretend to be in a different country. If the system uses a fairness control that applies corrections conditioned on these protected attributes, the fraudster may be able to tweak their attributes to maximize their success. Data poisoning attacks that target fairness controls have been recently developed [54,78].

Anomaly detection is also important to fraud detection for flagging novel and potentially malicious behaviors but has its own set of fairness pitfalls. Recent work [21,76] aims to quantify and mitigate these issues, but overall, fairness in anomaly detection systems is still a novel area of research.

Uncertainty quantification has clear applications in the settings of *active learning* (see Sect. 7) and *classification with a reject option* [18,33], where the ML model has the option to refuse or delay making a decision when uncertainty is too high around the model's prediction. Hüllermeier and Waegeman [36] provide a detailed survey of methods for *uncertainty quantification*.

Classification with a reject option has applications in fraud detection for so-called *trust systems*, which are tasked with assigning a permanent trust status to some subsets of the *genuine users* that are important customers and are clearly not fraudulent users, for which taking any action on them should at all times be avoided. Accidentally trusting fraudulent users could do a lot of damage, and therefore making a prediction can be rejected by the model when there is too much uncertainty regarding the instance.

An important concept in both *classification with a reject option* and *active learning* is *epistemic uncertainty*, i.e., the degree of uncertainty due to lack of data in the part of feature space for which the prediction needs to be made (reducible uncertainty). This contrasts *aleatoric uncertainty*, i.e., the degree of uncertainty due to an overlap in class distributions in the part of input space for which the prediction needs to be made (irreducible uncertainty). Methods that explicitly quantify the *epistemic* part of the uncertainty and can separate this from the *aleatoric* part are summarized in [36] and include *density estima-tion, anomaly detection, Bayesian models*, or the framework of *reliable classifica-tion* [73]. The argument is that rejecting or delaying a decision is only a reason-able decision if it is expected that the uncertainty is expected to *decrease*, which is not the case with aleatoric (irreducible) uncertainty. Likewise, in the active learning setting, spending the fraud investigator's time to investigate an instance only makes sense if there is reducible uncertainty regarding that instance. Mark-ing a user as trustworthy seems safe when a model predicts a user to be of the non-fraud class with *low epistemic uncertainty* regarding that prediction. How-ever, quantification of *epistemic uncertainty* is a rather novel research direction in the field of machine learning, and more research is needed. More specifically, the application of epistemic uncertainty quantification in adversarial problem domains is not well-understood.

Rules-based systems are created by *fraud investigators* and are designed to supplement ML models in the detection of fraud. Because of concept drift, rules can become ineffective shortly after they have been added to the *decision system*. After fraud investigators have first observed a new type of fraud attack that is not recognized by the ML model, a rule can be used for the period until the ML model picks up on that attack. In some sense, decision-making based on model output and multiple rule outputs can be seen as analogous to *ensemble learning* [69], which concerns the decision-making based on multiple ML models. A specific challenge is that the set of rules is subject to change: new rules get developed and old ones get decommissioned. This creates a need for ensemble models that combine a non-stationary set of components.

4.2   Open Research Challenges

Challenge 3: Model calibration under adversarial drift. Existing model calibration techniques help to generate calibrated probabilities on instances that are identically distributed as the calibration set. There are open ques-tions whether there are any bounds to the degree to which calibration can break down in an adversarial drift setting and whether this can be mitigated.

Challenge 4: Guidelines for design of cost-sensitive learning cost func-tions. There is a need for frameworks and guidelines for the engineering of cost functions for cost-sensitive learning applications in situations where not all relevant factors of the cost are easily expressed in financial value.

Challenge 5: Fairness and vulnerability. There are open questions regard-ing whether AI fairness methods could open up potential vulnerabilities in adversarial settings. Furthermore, there are open challenges around fairness in scenarios when the sensitive attributes can be spoofed by fraudsters.

Challenge 6: Fairness in anomaly detection. AI fairness in the area of anomaly detection systems is still a novel area of research. It is unclear if post-processing methods for AI fairness that work in the supervised setting are applicable in the setting of anomaly detection.

Challenge 7: Epistemic uncertainty quantification for trust systems.
Predictions of the non-fraud class that are made with *low epistemic uncertainty* could have an application to identify trustworthy users that should never be marked as fraudulent. However, more research is needed into the applications of such techniques in adversarial problem settings.

Challenge 8: Ensemble learning for non-stationary sets of components.
The output of the ML model and the rules ultimately need to be combined into a single decision. Ensemble learning methods address this task but do currently not handle dynamic sets of ensemble components.

5   Selection Bias in Labels

Labels are obtained through *fraud investigations* ⑩, and *automatic escalations* ⑫. From the *machine learning* perspective, we would like to train models using labeled instances that are uniformly sampled from the population. In practice, there are several sources of *selection bias*. First, *delay in labeling* arises for both label sources: manual investigations can take minutes up to days, whereas it can take days up to weeks for notifications of fraud to arrive through escalations [19]. Secondly, manual investigations may *overlook* fraudulent instances (e.g., caused by resource constraints, or well-hidden fraud). Third, *automated actions* ⑧ block suspicious transactions, and as a result, there will be merely a suspicion (no documented evidence) that these transactions are fraudulent. Because there is no certainty that these blocked transactions are fraudulent, they cannot be considered to be labeled instances. In many e-commerce applications, similar issues are commonly addressed with a *control group*, i.e., by always approving a certain percentage of transactions. While this control group would be an unbiased sample of labeled data, collecting it would come at the high cost of needing to purposefully let a share of fraud go through without blocking it.

5.1   Summary of the Literature

Learning under selection bias has been studied for example in [86] and [80], both relying on ideas from *causal inference* such as *inverse propensity weighting*. Another example worth mentioning is [35] where *matching* weights are computed directly. In [37], the authors study the problem in the low prevalence regime, which is particularly fitting the fraud detection problem, although they don't construct an unbiased model. Rather, they propose to utilize unlabeled data to construct a case ranking model, which might or might not be appropriate depending on the specific problem at hand. The *domain adaptation* field also tackles this problem, defined as learning a model with data sampled from a *source domain* to be applied in a *target domain*. This particular field distinguishes several settings, two of which are of particular interest to fraud detection:

Unsupervised domain adaptation (e.g., [26,79]) considers labeled and unlabeled examples from the source domain and unlabeled examples from the target domain, matching the no-control-group setting.

Semi-supervised domain adaptation (e.g., [20]) adds some labeled examples from the target domain, matching the with-control-group setting.

Multi-armed bandits (MAB) [43] is an area of research that studies the trade-off between *exploration* and *exploitation*. A control group is a simple form of exploration. In the context of automated actions, approving a transaction allows us to observe both the consequences of approving as well as the consequences that would have been observed if the transaction would have been blocked. This is known as *partial feedback* and is different from the *bandit feedback* setting where feedback is only observed for the action that was taken. No feedback is observed for blocked transactions.

To handle this setting, one can create more sophisticated exploration policies, that don't necessarily approve cases uniformly at random but explore with some optimization criteria. The principle of *optimism in the face of uncertainty* [7, 45,50] is of particular interest for the fraud detection problem. The core idea is to prioritize exploration (acceptance) of transactions where the expected cost is lower or the model has higher uncertainty. Typically, the expected cost is estimated through standard supervised learning techniques and the uncertainty is modeled with the variance of the mean cost estimate. Other approaches such as *Thompson sampling* [2,81], or more generally, *posterior sampling* [68] can balance exploration and exploitation in fraud detection.

5.2   Open Research Challenges

Challenge 9: Bias-variance trade-off. Removing the bias from the data almost always involves an increase in the variance of the predictions. This variance might lead to models with poor generalization error, defeating the purpose of bias reduction. Most bias reduction techniques focus on completely removing the bias, and although there exists work on variance reduction, it is always under the no-bias constraint. Creating principled mechanisms to tune this trade-off, potentially allowing positive bias but improving generalization error is still an open challenge, and an active area of research, mainly in the *domain adaptation* field. A related and harder challenge is the fact that in practice, at training time, there is no data available from the *target domain*, which can be considered an adversarial version of the *unsupervised domain adaptation* problem where the goal is to learn a model that generalizes *sufficiently well* to a large set of potential target domains from labeled source domain data.

Challenge 10: Pseudo-MAB setting. Only the approval action reveals full feedback whereas rejection reveals no feedback. This setting does not exactly match the MAB setting. This opens questions about the optimality of standard MAB policies. An alternative formulation is simply to select a subset of transactions for rejection (or acceptance) to minimize some carefully crafted loss function that combines the monetary costs with the value of the gathered information. This can be addressed from the perspective of set-function optimization and *online active learning* [71]. However, the MAB formulation addresses other relevant challenges such as delayed feedback and non stationary which have been studied to a large extent in the MAB literature (e.g., [30,38]).

6   Concept Drift

In Fig. 1, the cycles $\Rightarrow$(①, ②, ⑥, ⑦, ⑧) and $\Rightarrow$(①, ②, ⑥, ⑦, ⑨, ④) highlight how fraud detection is an adversarial problem domain. When the *decision system* is successful in blocking the fraud attempts of a fraudulent user (i.e., ④ or ⑧), then the fraudster is likely to try to circumvent the system by modifying their attack until successful. Due to this behavior, fraud detection systems often experience concept drift nearly constantly.

Besides adversarial drift from changing fraud attacks, the data distributions that are generated by genuine users can also be subject to concept drift. Examples include seasonal patterns, unexpected events (e.g., COVID-19), or changes in the e-commerce platform. However, the drift of genuine users is often largely independent of ④ and ⑧ and is thus not adversarial.

The third source of concept drift to fraud detection systems is related to updates to so-called *upstream* models. For example, imagine a webshop that requires users to log in before they can make a purchase. An update to a login-time fraud detection model shifts the distributions of the data that reaches a payment-time fraud detection model that occurs later in the sales funnel, because the population of fraudulent users that are already caught by the login-time model will likely change with the update.

6.1   Summary of the Literature

Concept drift adaptation [25,49], or *dataset shift* [64], is a well-studied research topic. Drift can be categorized by their **distributional type**: *covariate shift* concerns a shift in $P(X)$, *prior shift* a shift in $P(y)$, and *real concept drift* a shift in $P(y \mid X)$. Orthogonally, drift can be categorized by its **temporal type**: it can be *sudden, gradual, incremental,* or *recurring*. Finally, drift can be *adversarial* or *natural*, i.e., *adversarial* drift is specifically aimed to beat a detection system, while *natural* drift happens for reasons that are exogenous to it.

Fraud detection has adversarial drift in the *fraudulent* class. Changes to the attack patterns of fraudsters often result in gradual and incremental drift, as fraudsters tend to gradually increase the frequency of their successful and undetected attacks while decreasing the frequency of detected and unsuccessful attacks. Fraudsters can also cause recurring drift, as they occasionally retry old attempts to check whether the fraud detection systems still catch them.

Drift due to changes in the e-commerce platform is often *sudden* because changes (e.g., in an account registration portal or a payment process) change at once at the time of new code deployment. However, in practice, many changes in the e-commerce platform are first evaluated in an A/B test, and thus, the drift that results from this change might at first affect only a fraction of the users.

Furthermore, the drift that results from changes in the e-commerce platform is natural drift, contrasting the adversarial drift that originates from fraudsters' attempts to remain undetected. Much of the adversarial concept drift detection and adaptation literature ignores that such tasks often need to be performed *in the presence of sudden and natural drift* that originates from changes to the platform itself. While many concept drift detection techniques exist [25], there is a practical need for methods that can distinguish the fraudsters' gradual adversarial drift from the sudden and natural drift that is caused by platform changes.

Delayed labels make the task of concept drift adaptation much more challenging. Until the labels are known, concept drift is only detectable when a change in $P(y \mid X)$ is accompanied by a change in $P(X)$ [88]. Likewise, adaptation to a change in $P(y \mid X)$ is not possible without a change in $P(X)$. Several methods exist to address the problem of concept drift adaptation under delayed labels, including *positive unlabeled* (PU) learning [22,40], or by explicit modeling of the expected label delay of individual instances through survival modeling. Dal Pozzolo et al. [19] proposed a solution specific for the fraud detection case where they train two separate models. The first model is trained on the labels found by *fraud investigators*, while the second is trained on labels obtained through the often much more delayed label-source of escalations. In practice, some fraud detection use cases deal with label delay that is theoretically upper bounded, such as in the case of credit card chargebacks that have a deadline set by the credit card issuers. To the best of our knowledge, concept drift detection under *upper-bounded label delay* has not been studied as of yet.

Supervised methods for fraud detection often outperform purely unsupervised anomaly detection for fraud detection in industry applications [29]. Supervised models, in particular, outperform anomaly detection models in detecting fraud instances that are continuations of fraud attacks that were ongoing at the time the model was trained. The field of *evolving data stream classification* developed several methods to incrementally update ML models in a streaming setting to adapt to distributional changes. State-of-the-art methods include *adaptive random forest* [27] and *streaming random patches* [28]. The field is heavily focused on *updating* ML models instead of retraining them from scratch, which is motivated by computational efficiency. In practice, however, e-commerce organizations do have the computational resources that are required to retrain models daily.

Finally, there is currently limited insight into *how* fraudsters respond, adapt their attacks, and cause drift. In practice, the fraudsters don't have *direct* control over the feature vectors that their attacks produce, but they instead control it only *indirectly* through their interactions with the e-commerce platform. This constrains how fraudsters can change the distributions of feature values that they generate. The field currently lacks methods to identify potential fraud attacks that the e-commerce platform theoretically would allow for but that have not yet been observed. One possible direction is the use of *attack trees* [51], a common method in the cybersecurity field to map out possible attack angles for hackers.

Adversarial robustness in ML [11] is a research area that focuses on building ML models that make it hard for attackers to create *adversarial examples*, i.e., data points that the model predicts wrongly. Adversarial robustness closely links to concept drift–a fraud detection system that is adversarially robust makes it more difficult for fraudsters to generate new types of fraud that remain undetected. Current work on adversarial robustness is heavily focused on computer vision and natural language processing tasks, while the majority of fraud detection systems use tabular data. Adversarial robustness methods for tabular data are an open research challenge with applicability in fraud detection.

Anomaly detection is a class of methods that separate *normal* data points from *outlier* data points. The task of anomaly detection strongly links to *density estimation*, and can be seen as its inverse. *Novelty detection* [60] concerns the detection of *novel* behavior that emerges after drift and is therefore of particular relevance to fraud detection. Novelty detection typically uses anomaly detection– what is normal w.r.t. pre-drift data is likely to be an outlier w.r.t. post-drift data. Several empirical benchmark studies [1, 23] have compared anomaly detection methods, often identifying *isolation forest* [48] performs well consistently.

New attacks by fraudsters generate feature values that are distinct from their previous attacks, causing the new attacks to be marked as outliers. However, a drift in the behavior of genuine users (e.g., due to changes in the e-commerce platform or exceptional events like COVID-19) is also likely to generate feature values that are distinct from earlier behavior. Therefore, not every outlier can be assumed to be fraudulent, and not every distribution shift is caused by a change in fraud attacks. Marking all outliers as possible fraud cases that require investigation by the *fraud investigator* introduces spikes of false positives. In practice, such a spike would for example be expected with every release of a new change in the e-commerce platform. This calls for investigation into methods that account for the existence of "harmless" outliers caused by external changes.

6.2   Open Research Challenges

Challenge 11: Separating platform changes from changes in attack patterns. Changes in the e-commerce platform and changes in fraudster behavior both drive concept drift. In the former case, drift tends to be natural and sudden, while in the latter case it tends to be adversarial and gradual. Concept drift detection methods that alert in the latter case but not in the former would be of practical value for fraud detection.

Challenge 12: Accounting for platform changes in novelty detection. Changes in the e-commerce platform can cause large spikes in the number of outliers that are flagged by novelty detection algorithms, thereby limiting their practical use. This creates a need for methods that aim to detect outliers that are novel fraud types, but not outliers that result from platform changes.

Challenge 13: Mapping attack angles. There is a need for methods and frameworks to map possible attack-angles in an e-commerce platform, and for decision-making frameworks to leverage those this in work prioritization.

Challenge 14: Methods for adversarial robustness for tabular data. Many fraud detection systems work on tabular data, which is an understudied data modality in the research field of adversarial robustness.

Challenge 15: Balancing anomaly detection and supervised methods.
While *supervised methods* are more accurate in detecting recurring fraud
types, *anomaly detection methods* can detect new attacks. Can a strategy for
combining both types of models be automatically inferred?

Challenge 16: Concept drift adaptation in the delayed label setting.
Supervised methods for *concept drift adaptation* often assume that labels are
immediately available. How do we adapt to concept drift if labels may be
delayed? In some fraud problems, there is a theoretical upper bound in the
label delay. Can this upper bound be used in concept drift adaptation?

7 ML-Investigator Interaction

Fraud investigations are a vital part of the operational model by stopping fraudulent behavior through manual action ④, and as a result, generating new labels
for the ML model ⑩. Two objectives are involved here: the goal to *identify
fraud* and the goal to *generate labels that are most useful to the model*. These
two objectives can sometimes compete. Below we describe various machine learning (ML) methods to trigger investigations ⑨ in ways that address and balance
these objectives, and we discuss open research questions.

7.1 Summary of the Literature

Active learning (AL) is a paradigm that naturally fits cycle ⇒(⑥, ⑦, ⑨,
⑩) in Fig. 1. The AL paradigm decides which unlabeled data points to prioritize for labeling depending on how much they are expected to improve the ML
model. *Fraud investigators* label these data points, after which the model can
be retrained and a new iteration of data point prioritization is started. AL can
help to generate rapidly when escalations (i.e., ⑪ and ⑫) are slow. This helps
to mitigate label delay (see Sect. 5) and therefore improves the model's ability to
adapt to concept drift (see Sect. 6). Furthermore, selecting instances that would
maximize the learning is an efficient use of the fraud investigator's time.

AL has not been widely applied in industry context so far despite extensive
academic research [6,75]. This might be due to uncertainty about which AL
technique to use, how to deal with extreme class imbalance, the possibility of
viable alternatives, the engineering overhead, and uncertainty about the validity
of the assumptions made in AL. The problem of class imbalance is particularly
relevant for many fraud detection use cases. Carcillo et al. [14] investigated AL
methods under the high class-imbalance setting of fraud detection and showed
on credit card fraud data that simply selecting the instances with the highest
probability of being fraudulent maximizes learning and obtained high precision.
The popular uncertainty sampling [74] method explores those data points with
high proximity to the model's decision boundary. More recent work on active
learning [57] aims to distinguish *epistemic uncertainty* from *aleatoric uncertainty*
(see Sect. 4). The rationale is that the fraud investigator's investigation time is
wasted time if they investigate data points in parts of the feature space where

the uncertainty cannot be reduced (i.e., where uncertainty is aleatoric). *Inter-rater disagreement* between multiple fraud investigators about the same instance can cause aleatoric uncertainty. Traditional AL methods that do not distinguish between the two uncertainty types tend to repeatedly select instances with high *aleatoric* uncertainty [57], thereby wasting time of the fraud investigator. New fraud patterns are likely to come from regions of the feature space with high *epistemic* uncertainty. Therefore, fraud detection AL systems would benefit from focusing on sampling instances based on epistemic uncertainty. This research area is novel and lacks real-life evaluations in fraud settings.

Guided learning [5] contrasts by asking fraud investigators to search themselves for fraudulent examples (i.e., $\Rightarrow(\text{③}, \text{⑩})$), instead of being asked to provide labels for specific instances that were selected by the AL model. Guided learning is possible when investigators have sufficient domain knowledge to find positive examples themselves. This is typically the case in the fraud domain. Guided learning can be particularly useful when prevalence is very low and in situations with disjunct classes, like when there are different fraud modus operandi. A disadvantage is that the cost per label for guided learning is most likely higher than for AL. A further disadvantage is that relying on investigator searches induces selection bias that is unique to each investigator, the impact of which has not been studied to the best of our knowledge. Practically, guided learning can be supported by empowering investigators to generate queries based on ML models' input features [72]. This approach allows investigators to directly investigate specific areas of the feature space. While it is not a well-studied methodology, this presents a potential area of research.

Guided learning and AL can complement each other: fraud investigators can both proactively search for fraud and label suggestions from an AL model. Some success has been obtained with hybrid variants that start with guided learning and evolve to AL when some initial data set has been gathered [5], or that supplement AL with additional searched labels [10]. Guided learning is particularly successful compared to AL in settings with low prevalence [5]. However, the exact success factors in applications of searching and labeling are not well-understood beyond the dependence on prevalence.

Weak supervision techniques such as *snorkel* [65] solve the problem of inferring labels for instances using so-called *labeling functions* that *fraud investigators* create. These *labeling functions* are expected to be imperfect (i.e., *weak*), and can be seen as analogous to the *rules* in relation ⑤. The main idea of weak supervision is to infer *reliable labels* from a collection of weak labels using a generative model, which can then be used as ground truth to train the fraud detection model. An important aspect of weak supervision tools is a user interface that allows a *fraud investigator* and an ML practitioner to collaborate and develop new labeling functions that assign labels to instances that are not yet labeled by existing labeling functions. Fraud investigators must be able to quickly find patterns in currently unlabeled data points and develop new rules to apply weak supervision successfully. This procedure is represented by the cycle $\Rightarrow(\text{③}, \text{⑩})$ and by ⑤. In existing literature, like [65], the focus of this user interface is on textual data, where it is for example easy for a *fraud investigator* to instantly spot whether a tweet is *spam* or not. In practice, many fraud detection problems concern tabular data, and much more expert knowledge and deeper investigations are needed for the fraud investigator to conclude if a certain instance is fraudulent. This requires further research for user interfaces that support the iterative process of developing labeling functions for weak-supervision in the context of tabular data.

7.2   Open Research Challenges

Challenge 17: Exploration/exploitation trade-off in active learning. While in typical use cases of active learning the goal is to label data points that are most helpful for improving the model, in the fraud detection use case it is important to trade-off this goal with finding more fraud cases. The trade-off between these two goals is currently an open research challenge.

Challenge 18: Epistemic vs. aleatoric uncertainty sampling. How can we leverage active learning methods, while avoiding wasting our investigative resources on parts of the feature space with high aleatoric uncertainty? Epistemic uncertainty sampling is a promising direction of research, but applications of epistemic uncertainty estimates for active learning in a practical fraud prevention setting with adversarial drift are lacking.

Challenge 19: Label vs. search. The relative value of AL-type labeling and guided-learning-type searching depends on the cost of the two types of investigations and the fraud prevalence. In many situations, searching is most likely more expensive than labeling, but the exact conditions that influence the costs of both approaches are not clear.

Challenge 20: Weak supervision tools for fraud detection. The applicability of weak supervision methods is highly dependent on the ability to quickly and accurately assess the class of observed data points. This is often difficult because fraud investigations can be complex and time-consuming. Therefore, better decision support tools for fraud investigators are needed not only to assist the fraud investigations themselves but are also a requirement for applications of weak supervision methods for fraud detection.

8   Model Deployment and Monitoring

Training, deployment, and monitoring of ML models ⑥ in a production environment comes with a variety of challenges, some of which are specific to the setting of fraud detection. Fraud detection models are often integrated into vital parts of the e-commerce platform, such as the payment portal or account registration portal. The financial business consequences are large when such systems malfunction. For example, the business revenue would almost come to a complete halt if an outage would cause the platform to be unable to process payments or process login requests. Note that this situation is distinct from, for example, a recommender system, where an outage would be undesirable but of smaller consequences. Therefore, it is important to take risk mitigation measures to ensure that model deployment is safe. Additionally, fraud models are particularly retrained and deployed frequently compared to ML models in other parts of the business, because the adversarial drift creates a need to do so (see Sect. 6). This creates a need for deployment safety measures to be efficient and automated.

8.1   Summary of the Literature

Model verification methods allow to validate that the ML model satisfies certain desired properties. *CheckList* [67] is a verification method inspired by *metamorphic testing* in software engineering and requires the ML practitioner to formulate a set of unit-test-like checks that the model needs to pass. While the main use case of CheckList is during offline evaluation, these unit tests can additionally be used as a sanity check to verify that a model that has been deployed in the model serving platform indeed still passes the unit tests. For natural language data, CheckList provides a mechanism to automatically generate test cases at scale. For other forms of data, formulating test cases is currently still a manual process. Automated test case generation for data formats other than natural language is still an open challenge.

Deployment best practices have recently increasingly become a subject of study, and provide guidance on how to manage and mitigate the risks that are involved in model deployment. Early work includes the *ML test score* [12], which provides a list of checks to be performed during model deployment that can catch common problems and mistakes. More recent work includes [44,58]. Commonly recommended is the practice of *canary testing*, i.e., to expose a newly deployed model first to a small group of users where it is closely monitored before exposing all users to the model. Alternatively, in a *shadow mode* deployment, the new model starts making predictions for every instance without being used by the *decision system*. Another common recommendation is to create the infrastructure that allows for quick and safe rollbacks to an earlier model version, which enables quick recovery in case of unforeseen problems.

Automated data validation methods focus on monitoring and validation of the feature values that are used in the ML model. Fraud detection models often consume data generated by parts of the *e-commerce. platform* that are often not directly maintained by the anti-fraud department. For example, a payment processing service (or an accounts registration portal) often has a dedicated team that builds the service, owns its database tables, and controls its data schema. These database tables are then read by the fraud detection system to calculate feature values. There is a risk that newly deployed changes (or bugs) in these upstream dependencies affect the feature values and therefore the performance of the production model. For example, the accounts registration portal could contain an `age` field. A fraud detection model that uses this value as feature is negatively impacted when the team managing the registration portal changes the semantics of the age field by changing it from a mandatory to an optional field.

Automated data validation for ML applications has been studied extensively [15,62,70]. Solutions typically perform simple checks, such as validating that all feature values are within a reasonable range of values (e.g., age must not be negative), or by validating that the feature values have a reasonable distribution (e.g., values are not constant). Another frequent approach is to validate that *recent values* of features are within some threshold of similarity compared to *older values* of that same feature (e.g., using Kullback-Leibler divergence). While comparing feature distributions over time is useful to detect *change*, in practice, there is an important distinction between a change to *the semantics of a feature*[1] and *a change in user behavior*. The former case might require the data team to repair data pipelines, while the latter case requires the model to adapt to concept drift (see Sect. 6). There is an open challenge to detect changes in the semantics of features without false alarms on changes in user behavior.

8.2   Open Research Challenges

Challenge 21: Test case generation for model verification. There are existing methods for automated test case generation in the natural language domain. This is an open challenge for other types of data.

Challenge 22: Automated data validation under concept drift. Existing methods for automated data validation often compare recent feature values to older values. There is a need for automated data validation methods that distinguish between the case of broken data pipelines or changes to feature semantics on the one hand and a change of user behavior on the other hand. This would enable alerting only in the former scenario.

9   Conclusion

We presented an operational model of how an anti-fraud department in an e-commerce organization operates. We formulated a list of practical challenges related to fraud detection, and we derived a list of machine learning research topics that are practically relevant and applicable in anti-fraud departments by addressing some of these practical challenges. We summarized the state of the scientific literature in these research topics and formulated open research challenges that we believe to be relevant to the industry for anti-fraud operations.

By formulating these open challenges, this paper functions as a research agenda with industry practicality in mind. At the same time, this paper aims to enable future work in fraud detection to embed their methods in the *organizational context* using the operational model presented in this paper.

---

[1] Such as the age field that becomes optional. Or, alternatively, a temperature field that changes its implementation from Celsius to Fahrenheit.

References

1. Aggarwal, C.C., Sathe, S.: Outlier Ensembles: An Introduction. Springer, Cham (2017). https://doi.org/10.1007/978-3-319-54765-7
2. Agrawal, S., Goyal, N.: Analysis of Thompson sampling for the multi-armed bandit problem. In: Conference on Learning Theory, pp. 39.1–39.26. JMLR Workshop and Conference Proceedings (2012)
3. Amershi, S., Cakmak, M., Knox, W.B., Kulesza, T.: Power to the people: the role of humans in interactive machine learning. AI Mag. **35**, 105–120 (2014)
4. Amores, J.: Multiple instance classification: review, taxonomy and comparative study. Artif. Intell. **201**, 81–105 (2013)
5. Attenberg, J., Provost, F.: Why label when you can search?: Alternatives to active learning for applying human resources to build classification models under extreme class imbalance. In: Proceedings of the 16th ACM SIGKDD International Conference on Knowledge Discovery and Data Mining (2010)
6. Attenberg, J., Provost, F.: Inactive learning? Difficulties employing active learning in practice. SIGKDD Explor. Newsl. **12**(2), 36–41 (2011)
7. Auer, P.: Using confidence bounds for exploitation-exploration trade-offs. J. Mach. Learn. Res. **3**(Nov), 397–422 (2002)
8. Bansal, G., et al.: Does the whole exceed its parts? The effect of AI explanations on complementary team performance. In: Proceedings of the 2021 CHI Conference on Human Factors in Computing Systems (2020)
9. Beutel, A., et al.: Putting fairness principles into practice: challenges, metrics, and improvements. In: Proceedings of the 2019 AAAI/ACM Conference on AI, Ethics, and Society (2019)
10. Beygelzimer, A., Hsu, D.J., Langford, J., Zhang, C.: Search improves label for active learning. In: Lee, D., Sugiyama, M., Luxburg, U., Guyon, I., Garnett, R. (eds.) Advances in Neural Information Processing Systems, vol. 29. Curran Associates, Inc. (2016)
11. Biggio, B., Roli, F.: Wild patterns: ten years after the rise of adversarial machine learning. Pattern Recogn. **84**, 317–331 (2018)
12. Breck, E., Cai, S., Nielsen, E., Salib, M., Sculley, D.: The ML test score: a rubric for ml production readiness and technical debt reduction. In: 2017 IEEE International Conference on Big Data (Big Data), pp. 1123–1132. IEEE (2017)
13. Carbonneau, M.A., Cheplygina, V., Granger, E., Gagnon, G.: Multiple instance learning: a survey of problem characteristics and applications. CoRR abs/1612.03365 (2016)
14. Carcillo, F., Borgne, Y.L., Caelen, O., Bontempi, G.: Streaming active learning strategies for real-life credit card fraud detection: assessment and visualization. Int. J. Data Sci. Anal. **5**(4), 285–300 (2018)
15. Caveness, E., Paul Suganthan, G.C., Peng, Z., Polyzotis, N., Roy, S., Zinkevich, M.: TensorFlow data validation: data analysis and validation in continuous ML pipelines. In: Proceedings of the 2020 ACM SIGMOD International Conference on Management of Data, pp. 2793–2796 (2020)
16. Cavico, F.J., Mujtaba, B.G.: Wells Fargo's fake accounts scandal and its legal and ethical implications for management. SAM Adv. Manag. J. **82**(2), 4 (2017)
17. Chouldechova, A., Prado, D.B., Fialko, O., Vaithianathan, R.: A case study of algorithm-assisted decision making in child maltreatment hotline screening decisions. In: FAT (2018)

18. Chow, C.: On optimum recognition error and reject tradeoff. IEEE Trans. Inf. Theory **16**(1), 41–46 (1970)
19. Dal Pozzolo, A., Boracchi, G., Caelen, O., Alippi, C., Bontempi, G.: Credit card fraud detection and concept-drift adaptation with delayed supervised information. In: 2015 International Joint Conference on Neural Networks (IJCNN), pp. 1–8. IEEE (2015)
20. Daumé, H., III, Kumar, A., Saha, A.: Frustratingly easy semi-supervised domain adaptation. In: Proceedings of the 2010 Workshop on Domain Adaptation for Natural Language Processing, pp. 53–59 (2010)
21. Davidson, I., Ravi, S.: A framework for determining the fairness of outlier detection. In: ECAI (2020)
22. Elkan, C., Noto, K.: Learning classifiers from only positive and unlabeled data, pp. 213–220 (2008)
23. Emmott, A., Das, S., Dietterich, T., Fern, A., Wong, W.K.: A meta-analysis of the anomaly detection problem. arXiv preprint arXiv:1503.01158 (2015)
24. Foulds, J.R., Frank, E.: A review of multi-instance learning assumptions. Knowl. Eng. Rev. **25**(1), 1–25 (2010)
25. Gama, J., Žliobaitė, I., Bifet, A., Pechenizkiy, M., Bouchachia, A.: A survey on concept drift adaptation. ACM Comput. Surv. (CSUR) **46**(4), 1–37 (2014)
26. Ganin, Y., Lempitsky, V.: Unsupervised domain adaptation by backpropagation. In: Bach, F., Blei, D. (eds.) Proceedings of the 32nd International Conference on Machine Learning. Proceedings of Machine Learning Research, Lille, France, 07–09 July 2015, vol. 37, pp. 1180–1189. PMLR (2015)
27. Gomes, H.M., et al.: Adaptive random forests for evolving data stream classification. Mach. Learn. **106**(9), 1469–1495 (2017)
28. Gomes, H.M., Read, J., Bifet, A.: Streaming random patches for evolving data stream classification. In: 2019 IEEE International Conference on Data Mining (ICDM), pp. 240–249. IEEE (2019)
29. Görnitz, N., Kloft, M., Rieck, K., Brefeld, U.: Toward supervised anomaly detection. J. Artif. Intell. Res. **46**, 235–262 (2013)
30. György, A., Joulani, P.: Adapting to delays and data in adversarial multi-armed bandits. arXiv preprint arXiv:2010.06022 (2020)
31. Hamilton, W.L.: Graph representation learning. Synth. Lect. Artif. Intell. Mach. Learn. **14**(3), 1 159 (2020)
32. Hardt, M., Price, E., Srebro, N.: Equality of opportunity in supervised learning. In: NIPS (2016)
33. Hellman, M.E.: The nearest neighbor classification rule with a reject option. IEEE Trans. Syst. Sci. Cybern. **6**(3), 179–185 (1970)
34. Holstein, K., Vaughan, J.W., Daumé, H., Dudík, M., Wallach, H.: Improving fairness in machine learning systems: what do industry practitioners need? In: Proceedings of the 2019 CHI Conference on Human Factors in Computing Systems (2019)
35. Huang, J., Gretton, A., Borgwardt, K., Schölkopf, B., Smola, A.: Correcting sample selection bias by unlabeled data. In: Schölkopf, B., Platt, J., Hoffman, T. (eds.) Advances in Neural Information Processing Systems, vol. 19. MIT Press (2007)
36. Hüllermeier, E., Waegeman, W.: Aleatoric and epistemic uncertainty in machine learning: an introduction to concepts and methods. Mach. Learn. **110**(3), 457–506 (2021)
37. Jacobusse, G., Veenman, C.: On selection bias with imbalanced classes. In: Calders, T., Ceci, M., Malerba, D. (eds.) DS 2016. LNCS (LNAI), vol. 9956, pp. 325–340. Springer, Cham (2016). https://doi.org/10.1007/978-3-319-46307-0_21

38. Joulani, P., Gyorgy, A., Szepesvári, C.: Online learning under delayed feedback. In: International Conference on Machine Learning, pp. 1453–1461. PMLR (2013)

39. Kawase, R., Diana, F., Czeladka, M., Schüler, M., Faust, M.: Internet fraud: the case of account takeover in online marketplace. In: Proceedings of the 30th ACM Conference on Hypertext and Social Media, pp. 181–190 (2019)

40. Kiryo, R., Niu, G., du Plessis, M.C., Sugiyama, M.: Positive-unlabeled learning with non-negative risk estimator. In: Proceedings of the 31st International Conference on Neural Information Processing Systems, pp. 1674–1684 (2017)

41. Kotsiantis, S., Kanellopoulos, D.: Multi-instance learning for predicting fraudulent financial statements. In: Proceedings of the International Conference on Convergence Information Technology, vol. 1, pp. 448–452. IEEE Computer Society (2008)

42. Kull, M., Silva Filho, T., Flach, P.: Beta calibration: a well-founded and easily implemented improvement on logistic calibration for binary classifiers. In: Artificial Intelligence and Statistics, pp. 623–631. PMLR (2017)

43. Lattimore, T., Szepesvári, C.: Bandit Algorithms. Cambridge University Press, Cambridge (2020)

44. Lavin, A., et al.: Technology readiness levels for machine learning systems. arXiv preprint arXiv:2101.03989 (2021)

45. Li, L., Chu, W., Langford, J., Schapire, R.E.: A contextual-bandit approach to personalized news article recommendation. In: Proceedings of the 19th International Conference on World Wide Web, pp. 661–670 (2010)

46. Liang, C., et al.: Uncovering insurance fraud conspiracy with network learning. In: Proceedings of the 42nd International ACM SIGIR Conference on Research and Development in Information Retrieval, pp. 1181–1184 (2019)

47. Ling, C.X., Sheng, V.S.: Cost-sensitive learning. In: Sammut, C., Webb, G.I. (eds.) Encyclopedia of Machine Learning, pp. 231–235. Springer, Boston (2010). https://doi.org/10.1007/978-0-387-30164-8_181

48. Liu, F.T., Ting, K.M., Zhou, Z.H.: Isolation forest. In: Proceedings of the 8th IEEE International Conference on Data Mining (ICDM), pp. 413–422. IEEE (2008)

49. Lu, J., Liu, A., Dong, F., Gu, F., Gama, J., Zhang, G.: Learning under concept drift: a review. IEEE Trans. Knowl. Data Eng. **31**(12), 2346–2363 (2018)

50. Mahajan, D.K., Rastogi, R., Tiwari, C., Mitra, A.: LogUCB: an explore-exploit algorithm for comments recommendation. In: Proceedings of the 21st ACM International Conference on Information and Knowledge Management, pp. 6–15 (2012)

51. Mauw, S., Oostdijk, M.: Foundations of attack trees. In: Won, D.H., Kim, S. (eds.) ICISC 2005. LNCS, vol. 3935, pp. 186–198. Springer, Heidelberg (2006). https://doi.org/10.1007/11734727_17

52. McAfee: The economic impact of cybercrime - no slowing down (2018). https://www.mcafee.com/enterprise/en-us/assets/executive-summaries/es-economic-impact-cybercrime.pdf. Accessed 05 June 2021

53. Mehrabi, N., Morstatter, F., Saxena, N., Lerman, K., Galstyan, A.: A survey on bias and fairness in machine learning. arXiv:1908.09635 (2019)

54. Mehrabi, N., Naveed, M., Morstatter, F., Galstyan, A.: Exacerbating algorithmic bias through fairness attacks. arXiv:2012.08723 (2020)

55. Moagar-Poladian, S., Dumitrescu, G.C., Tanase, I.A.: Retail e-commerce (e-tail)-evolution, characteristics and perspectives in China, the USA and Europe. Glob. Econ. Obs. **5**(1), 167 (2017)

56. Nanduri, J., Jia, Y., Oka, A., Beaver, J., Liu, Y.W.: Microsoft uses machine learning and optimization to reduce e-commerce fraud. INFORMS J. Appl. Anal. **50**(1), 64–79 (2020)

57. Nguyen, V.-L., Destercke, S., Hüllermeier, E.: Epistemic uncertainty sampling. In: Kralj Novak, P., Šmuc, T., Džeroski, S. (eds.) DS 2019. LNCS (LNAI), vol. 11828, pp. 72–86. Springer, Cham (2019). https://doi.org/10.1007/978-3-030-33778-0_7

58. Paleyes, A., Urma, R.G., Lawrence, N.D.: Challenges in deploying machine learning: a survey of case studies. arXiv preprint arXiv:2011.09926 (2020)

59. Pevný, T., Dedic, M.: Nested multiple instance learning in modelling of HTTP network traffic. CoRR abs/2002.04059 (2020)

60. Pimentel, M.A., Clifton, D.A., Clifton, L., Tarassenko, L.: A review of novelty detection. Signal Process. **99**, 215–249 (2014)

61. Platt, J., et al.: Probabilistic outputs for support vector machines and comparisons to regularized likelihood methods. Adv. Large Margin Classif. **10**(3), 61–74 (1999)

62. Polyzotis, N., Zinkevich, M., Roy, S., Breck, E., Whang, S.: Data validation for machine learning. Proc. Mach. Learn. Syst. **1**, 334–347 (2019)

63. Porwal, U., Mukund, S.: Credit card fraud detection in e-commerce. In: 2019 18th IEEE International Conference on Trust, Security and Privacy in Computing and Communications/13th IEEE International Conference on Big Data Science and Engineering (TrustCom/BigDataSE), pp. 280–287. IEEE (2019)

64. Quiñonero-Candela, J., Sugiyama, M., Lawrence, N.D., Schwaighofer, A.: Dataset Shift in Machine Learning. MIT Press, Cambridge (2009)

65. Ratner, A., Bach, S.H., Ehrenberg, H., Fries, J., Wu, S., Ré, C.: Snorkel: rapid training data creation with weak supervision. VLDB J. **29**(2), 709–730 (2020)

66. Ribeiro, M.T., Singh, S., Guestrin, C.: Anchors: High-precision model-agnostic explanations. In: AAAI (2018)

67. Ribeiro, M.T., Wu, T., Guestrin, C., Singh, S.: Beyond accuracy: behavioral testing of NLP models with checklist. In: Proceedings of the 58th Annual Meeting of the Association for Computational Linguistics (ACL), pp. 4902–4912 (2020)

68. Russo, D., Van Roy, B.: Learning to optimize via posterior sampling. Math. Oper. Res. **39**(4), 1221–1243 (2014)

69. Sagi, O., Rokach, L.: Ensemble learning: a survey. Data Min. Knowl. Discov. **8**(4), e1249 (2018)

70. Schelter, S., Lange, D., Schmidt, P., Celikel, M., Biessmann, F., Grafberger, A.: Automating large-scale data quality verification. Proc. VLDB Endow. **11**(12), 1781–1794 (2018)

71. Sculley, D.: Online active learning methods for fast label-efficient spam filtering. In: CEAS, vol. 7, p. 143 (2007)

72. Sculley, D., Otey, M.E., Pohl, M., Spitznagel, B., Hainsworth, J., Zhou, Y.: Detecting adversarial advertisements in the wild. In: KDD (2011)

73. Senge, R., et al.: Reliable classification: learning classifiers that distinguish aleatoric and epistemic uncertainty. Inf. Sci. **255**, 16–29 (2014)

74. Settles, B.: Active learning literature survey, July 2010

75. Settles, B.: From theories to queries: active learning in practice. In: Guyon, I., Cawley, G., Dror, G., Lemaire, V., Statnikov, A. (eds.) Active Learning and Experimental Design Workshop in Conjunction with AISTATS 2010. Proceedings of Machine Learning Research, Sardinia, Italy, 16 May 2011, vol. 16, pp. 1–18. JMLR Workshop and Conference Proceedings (2011)

76. Shekhar, S., Shah, N., Akoglu, L.: FairOD: fairness-aware outlier detection. arXiv:2012.03063 (2020)

77. Sheng, V.S., Ling, C.X.: Thresholding for making classifiers cost-sensitive. In: AAAI, vol. 6, pp. 476–481 (2006)

78. Solans, D., Biggio, B., Castillo, C.: Poisoning attacks on algorithmic fairness. In: ECML/PKDD (2020)

79. Sun, B., Feng, J., Saenko, K.: Return of frustratingly easy domain adaptation. In: Proceedings of the Thirtieth AAAI Conference on Artificial Intelligence, AAAI 2016, pp. 2058–2065. AAAI Press (2016)

80. Swaminathan, A., Joachims, T.: Counterfactual risk minimization: learning from logged bandit feedback. In: Bach, F., Blei, D. (eds.) Proceedings of the 32nd International Conference on Machine Learning. Proceedings of Machine Learning Research, Lille, France, 07–09 July 2015, vol. 37, pp. 814–823. PMLR (2015)

81. Thompson, W.R.: On the likelihood that one unknown probability exceeds another in view of the evidence of two samples. Biometrika **25**(3/4), 285–294 (1933)

82. Ustun, B., Liu, Y., Parkes, D.: Fairness without harm: decoupled classifiers with preference guarantees. In: Chaudhuri, K., Salakhutdinov, R. (eds.) Proceedings of the 36th International Conference on Machine Learning. Proceedings of Machine Learning Research, 09–15 June 2019, vol. 97, pp. 6373–6382. PMLR (2019)

83. Wang, R., Nie, K., Wang, T., Yang, Y., Long, B.: Deep learning for anomaly detection. In: Proceedings of the 13th International Conference on Web Search and Data Mining, pp. 894–896 (2020)

84. Weerts, H.J.P., van Ipenburg, W., Pechenizkiy, M.: A human-grounded evaluation of SHAP for alert processing. In: Proceedings of the KDD Workshop on Explainable AI (KDD-XAI) (2019)

85. Xiao, C., Freeman, D., Hwa, T.: Detecting clusters of fake accounts in online social networks. In: Proceedings of the 8th ACM Workshop on Artificial Intelligence and Security, pp. 91–101, October 2015

86. Zadrozny, B.: Learning and evaluating classifiers under sample selection bias. In: Proceedings of the Twenty-First International Conference on Machine Learning, p. 114. ACM (2004)

87. Zadrozny, B., Elkan, C.: Transforming classifier scores into accurate multiclass probability estimates. In: Proceedings of the Eighth ACM SIGKDD International Conference on Knowledge Discovery and Data Mining, pp. 694–699 (2002)

88. Žliobaite, I.: Change with delayed labeling: when is it detectable? In: 2010 IEEE International Conference on Data Mining Workshops, pp. 843–850. IEEE (2010)

Few-Sample Named Entity Recognition for Security Vulnerability Reports by Fine-Tuning Pre-trained Language Models

Guanqun Yang, Shay Dineen, Zhipeng Lin, and Xueqing Liu[✉]

Department of Computer Science, Stevens Institute of Technology,
Hoboken, NJ 07030, USA
{gyang16,sdineen1,zlin30,xliu127}@stevens.edu

Abstract. Public security vulnerability reports (e.g., CVE reports) play an important role in the maintenance of computer and network systems. Security companies and administrators rely on information from these reports to prioritize tasks on developing and deploying patches to their customers. Since these reports are unstructured texts, automatic information extraction (IE) can help scale up the processing by converting the unstructured reports to structured forms, e.g., software names and versions [8] and vulnerability types [38]. Existing works on automated IE for security vulnerability reports often rely on a large number of labeled training samples [8,18,48]. However, creating massive labeled training set is both expensive and time consuming. In this work, for the first time, we propose to investigate this problem where only a small number of labeled training samples are available. In particular, we investigate the performance of fine-tuning several state-of-the-art pre-trained language models on our small training dataset. The results show that with pre-trained language models and carefully tuned hyperparameters, we have reached or slightly outperformed the state-of-the-art system [8] on this task. Consistent with previous two-step process of first fine-tuning on main category and then transfer learning to others as in [7], if otherwise following our proposed approach, the number of required labeled samples substantially decrease in both stages: 90% reduction in fine-tuning from 5758 to 576, and 88.8% reduction in transfer learning with 64 labeled samples per category. Our experiments thus demonstrate the effectiveness of few-sample learning on NER for security vulnerability report. This result opens up multiple research opportunities for few-sample learning for security vulnerability reports, which is discussed in the paper. Our implementation for few-sample vulnerability entity tagger in security reports could be found at https://github.com/guanqun-yang/FewVulnerability.

Keywords: Software vulnerability identification · Few-sample named entity recognition · Public security reports

© Springer Nature Switzerland AG 2021
G. Wang et al. (Eds.): MLHat 2021, CCIS 1482, pp. 55–78, 2021.
https://doi.org/10.1007/978-3-030-87839-9_3

1   Introduction

Security vulnerabilities pose great challenges to software and network systems. For example, there are numerous reported data breaches of Uber, Equifax, Marriott, and Facebook that jeopardize hundreds of millions of customers' information security [15]; vulnerable in-flight automation software used in Boeing 737-MAX are found to be guilty for a series of crashes in 2018 [31]. To allow the tracking and management of security vulnerabilities, people have created public security vulnerability reports and store them in vulnerability databases for reference. For example, the Common Vulnerabilities and Exposures (CVE) database is created and maintained by MITRE[1]; the National Vulnerability Database (NVD) is another database maintained by the U.S. government. These databases have largely facilitated security companies and vendors to manage and prioritize deployment of patches.

To scale up the management of vulnerability entries, existing works have leveraged natural language processing (NLP) techniques to extract information from unstructured vulnerability reports. For example, software name, software version, and vulnerability type. The extraction of such information can help accelerate security administration in the following scenarios: First, after converting the unstructured reports into structured entries such as software name and software version, such information can be directly used for detecting the inconsistencies between different records of vulnerability reports. For example, significant inconsistency was detected between reports from the CVE and NVD databases [8], which may cause system administrators to retrieve outdated or incorrect security alerts, exposing systems under their watch to hazard; inconsistency was also detected between two entries created by the same vendor Samsung in [10]; Second, the extracted entries can be used in other downstream applications. For example, we can leverage different categories of entries (e.g., vendor name, vulnerability types, attacker name) to construct a knowledge ontology for modeling the interplay between security entries [2,16]. The extracted entries have also been used for automatically generating a natural language summarization of the original report by leveraging a template [38].

The scalable processing of vulnerability reports into structured forms relies on the support from an effective information extraction (IE) system. With the massive amount of reports, manual extraction of information is intractable and most existing system relies on machine learning approaches. For example, Mulwad et al. [30] built an SVM-based system which extracted security concepts from web texts. In the recent years, with the rise of deep neural networks, systems based on deep learning were proposed, e.g., by using LSTM [18,48] and bi-directional GRU [8]. Deep learning has been shown to be effective in this task, e.g., with an F1 score of 98% [8].

Nevertheless, a major challenge of applying existing systems (e.g., recurrent neural network like GRU [8] and LSTM [18,48]) is that they often require a large number of labeled training samples to perform well; this number could range from

---

[1] https://www.mitre.org/.

a few thousand [47] to tens of thousand [18]. Labeling a NER dataset specific to computer security domain at this scale is both costly and time-consuming. This problem of annotation is further exacerbated by the fact that new security reports are routinely generated. Therefore, it is vital to investigate the possibility to alleviate the burden of labeling huge corpus, i.e., few-sample learning.

In this paper, for the first time, we investigate the problem of information extraction for public vulnerability reports under the setting where only a small number of labeled training samples are available. Our goal is to minimize the number of labeled samples without suffering from a significant performance degradation.

In recent years, significant progress has been made for few-shot learning, e.g., image classification [12], text classification [17], text generation [3], etc. In these tasks, the training set and validation set often only contain ten or fewer examples, and it has been shown that the available small number of examples are all what the model needs to achieve a good performance [5]. To generalize to unseen examples, few-shot learning approaches often leverage meta-learning [12], metric learning [6,36] and model pre-training. For example, the GPT-3 model [3] contains 175 billion parameters and was pre-trained on 45TB of unlabeled text data. It is able to accomplish a variety of generative tasks (e.g., question answering, conversation generation, text to command) when the user only provides a few examples for demonstration. For information extraction, more specifically, named entity recognition (NER), several existing works study its few-shot setting, where only tens of labeled samples are provided during training [21,44]. However, NER systems to date can only achieve a 62% F1 score (evaluated on the CoNLL 2003 dataset [39]). When applied to the security vulnerability reports domain, it is thus unclear whether the performance can be satisfactory enough to security administrators and vendors, as a lower error rate is often required [8]. As a result, in this paper, we investigate the following research questions:

- Is it possible to match the state-of-the-art performance in NER for security vulnerability reports by using only a small number of labeled examples?
- What is the minimum number of labeled examples required for this problem?

To answer these research questions, we conduct an experimental study on *few-sample* named entity recognition for vulnerability reports[2], which to the best of our knowledge has not been explored by previous works. As an initial study, we focus on the VIEM dataset [8], a dataset of vulnerability reports containing 3 types of entities (i.e., *software name, software version, outside*) and 13 sub-datasets based on the category of vulnerabilities. For the largest sub-dataset, we find that through fine-tuning pre-trained language models, it is possible to match the state-of-the-art performance (i.e., 0.92 to 0.93 F1 score for different entity types [8]) using only 576 labeled sentences for the training dataset, or 10% of the

---

[2] Here we frame our problem under the term "few-sample" learning instead of "few-shot" learning because our approach generally requires tens to a few hundred labeled training samples, which is more than "few-shot". We adopt the name "few-sample" from [47].

original training data. For the other 12 categories, the same approach matches the SOTA performance with only 11.2% of the labeled samples. Our experiments also show that the simple fine-tuning approach works better than state-of-the-art systems for few-shot NER [44].

The main contributions of this paper are:

- We propose, for the first time, to study the problem of named entity recognition (NER) for security vulnerability reports;
- We perform an experimental study by fine-tuning three state-of-the-art language models and evaluating one metric learning-based method for few-sample NER in vulnerability reports, which shows that the fine-tuning approach can achieve the similar performance using only 10% or 11.2% of the original training dataset;
- We discuss multiple future research opportunities of information extraction for security vulnerability reports;

The rest of this paper is organized as follows. Section 2 defines the problem and discusses the domain and data-specific challenges; Sect. 3 introduces the methods examined; Sect. 4 describes the experimental steps, datasets, and experimental results; Sect. 5 analyzes the related work; finally, Sect. 6 draws the conclusion, analyzes the limitation of this work and discusses several future directions of research for few-sample learning for security vulnerability reports.

2 Problem Definition and Challenges

2.1 Few-Sample Named Entity Recognition

Name entity recognition (NER) is the task of assigning predefined entity names to tokens in the input text. Formally, with predefined tagging set $\mathcal{Y}$ and a sequence of tokens (input sentence) $X_i = [x_{i1}, x_{i2}, \cdots, x_{iT}]$, each x_{ik} ($k = 1, 2, \cdots, T$) corresponds to a tag $y_{ik} \in \mathcal{Y}$ specifying the its entity type. This gives X_i a tagging sequence $Y_i = [y_{i1}, y_{i2}, \cdots, y_{iT}]$ and a typical dataset of NER is of the form $\mathcal{D} = \{(X_i, Y_i)\}_{i=1}^N$.

Notice there exist two distinct definitions for what a few-shot learning algorithm tries to achieve [41]. In the first definition, the goal is to match the performance when only a subset of training data is available; in the second definition, the test set contains unseen classes in the training set and the algorithm needs to generalize to the unseen classes. In this work, we focus on the first setting. We refer readers interested in the details of these two settings to [47].

2.2 Named Entity Recognition for Vulnerability Reports

An example of security vulnerability report (CVE-2015-2384) is shown in Fig. 1, which is a snapshot from the NVD website[3]. In this paper, we focus on the VIEM

---

[3] https://nvd.nist.gov.

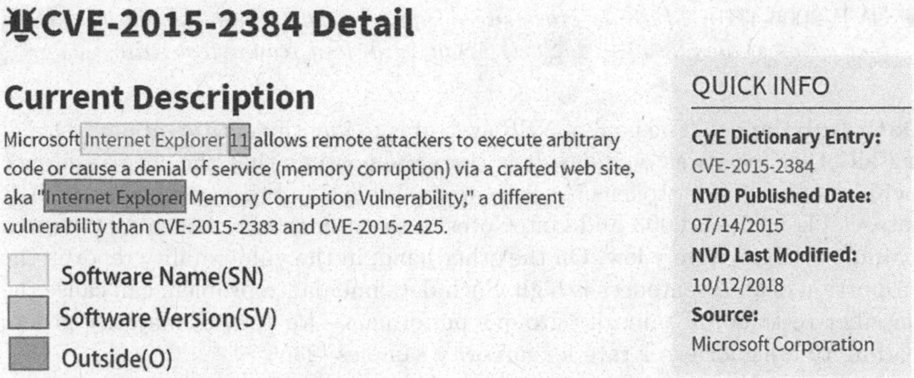

⚑CVE-2015-2384 Detail

Current Description

Microsoft Internet Explorer 11 allows remote attackers to execute arbitrary code or cause a denial of service (memory corruption) via a crafted web site, aka "Internet Explorer Memory Corruption Vulnerability," a different vulnerability than CVE-2015-2383 and CVE-2015-2425.

☐ Software Name(SN)

■ Software Version(SV)

■ Outside(O)

QUICK INFO

CVE Dictionary Entry:
CVE-2015-2384
NVD Published Date:
07/14/2015
NVD Last Modified:
10/12/2018
Source:
Microsoft Corporation

Fig. 1. An example of public security vulnerability report

dataset from [8], which contains 13 categories of vulnerabilities and two types of named entities: software name and software version. That is, the tag set $\mathcal{Y}$ is instantiated as $\{\text{SN}, \text{SV}, \text{O}\}$, where SN, SV, O refer to software name, software version, and non-entities, respectively. For example, as is shown in Fig. 1, the report CVE-2015-2384 warrants a security alert where the software **Internet Explorer** of version 11 might expose its vulnerability to attackers. The training, validation, and testing set share the same tag set $\mathcal{Y}$.

2.3   Data-Specific Challenges

There exists several data-specific challenges in the VIEM dataset, which we summarize as below.

Contextual Dependency. When identifying vulnerable software names and versions in vulnerability reports, the tagger has to take the entire context into account. For example, as is shown in Fig. 1 (CVE-2015-2384), **Internet Explorer** is tagged as a software name for its first occurrence; in the second occurrence, it appears in the vulnerability name *Internet Explorer Memory Corruption Vulnerabilities* thus is no longer a software name. Software version tag is also context dependent: in the examples below, 1.0 is tagged SV in CVE-2002-2323 and CVE-2010-2487 but O in CVE-2006-4710, this is because when the software name (James M. Snell) is tagged O, the software version should also be tagged O.

- CVE-2002-2323: *Sun PC NetLink* 1.0 *through 1.2 ... access restriction.*
- CVE-2010-2487: *Directory traversal vulnerability in YelloSoft Pinky* 1.0 *... in the URL.*

- CVE-2006-4710: *Multiple cross-site scripting ... as demonstrated by certain test cases of the* ⟨James M. Snell Atom⟩ ⟨1.0⟩ *feed reader test suite.*

Data Imbalance. The goal of NER system is to identify entities of interest and tag all other tokens as outsider. It is therefore possible that the entire sentence includes only outsider tokens. As is shown in Table 1, in the general domain NER dataset like CoNLL 2003 and OntoNotes 5, the proportion of such sentences in training set is relatively low. On the other hand, in the vulnerability reports, the proportion of all-O sentences is high. Such data imbalance problem can cause the classifier to trade off minority group's performance for that of majority group, leading to a higher error rate for minority samples [23].

Table 1. The proportion of sentences in training set that includes only non-entities. The statistics of CoNLL 2003 and OntoNotes 5 are estimated from publicly available dataset releases. In the VIEM dataset, the vulnerability category memc is intentionally balanced in [8] to ease transfer learning, making the aggregate (i.e. VIEM (all)) non-entity-only proportion lower.

Dataset	Domain	Non-entity-only sentences
CoNLL 2003[a]	General	20.71%
OntoNotes 5[b]	General	51.60%
VIEM (all)	Security	60.34%
VIEM (w/o memc)	Security	75.44%

[a]https://huggingface.co/datasets/conll2003
[b]https://github.com/Fritz449/ProtoNER/tree/master/ontonotes

3 Few-Sample NER for Vulnerability Reports

In this section, we introduce the methods investigated in this paper. The first method leverages fine-tuning pre-trained language models (PLM); the second method is a few-shot NER method based on the nearest neighbor approach [44].

3.1 Fine-Tuning Pre-trained Language Models with Hundreds of Training Labels

Fine-Tuning Framework and Selection of Pre-trained Language Models. Before we describe our approach, first we introduce our choices for the language model to be fine-tuned. There exists hundreds of candidate models from the HugginFace model hub[4]. We investigate three models out of all: bert-large-cased, roberta-large, and electra-base-discriminator. BERT was the first pre-trained language model that achieved great performance

---

[4] https://huggingface.co/models.

compared to prior art; RoBERTa and Electra are among the top-ranked language models on the GLUE leaderboard as of Jan 2021[5]. For BERT, we choose the cased version `bert-large-cased` rather than the uncased counterpart (i.e. `bert-large-uncased`) as token casings could be informative for the identification of software names (`SN`). Our implementation is based on the HuggingFace `transformers` library (4.5.1), more specifically, we leverage the `AutoModelForTokenClassification` class for fine-tuning the 13 categories in the `VIEW` dataset.

Fine-Tuning on the `memc` Category. Following the experimental protocol of VIEM [8], the PLM in our system is first fine-tuned on the vulnerability category of `memc` (short for "memory corruption") as, from the statistics of the dataset shown in Table 2, this category suffers less from data imbalance described in Sect. 2; fine-tuning on this category could therefore best help PLMs learn representations specific to computer security domain. However, rather than allowing PLMs accessing the entire training dataset of `memc`, we constrain its access to the training data from 1% to 10% of the entire training set.

Transfer Learning to the Other 12 Categories. Previous works on fine-tuning language models show that fine-tuning can benefit from multi-task learning or transfer learning, i.e., fine-tuning on a related task before fine-tuning on the objective task [34]. We adopt this approach, i.e., the PLM is first fine-tuned on the `memc` category, then we start from the checkpoint and continue the fine-tuning on the 12 categories. We again limit the number of annotated samples during transfer learning. Throughout the transfer learning experiment, we sample varying number of training samples from each of the 12 categories, and we combine the 12 subsets to form an *aggregate* training set. For validation and testing, however, we evaluate the model with transfer learning on each category *independently*.

Hyperparameter Optimization. The hyperparameters for fine-tuning includes the learning rate, batch size, and the number of training epochs. Inappropriately setting one of these hyperparameter might cause performance degradation [28]. As a result, hyperparameter tuning is critical as they could have substantial impact on the model performance.

Setting the Random Seed. The fine-tuning performance for PLM are known to be unstable, especially under the typical few-sample setting. Some seemingly benign parameters, like choice of random seed, could introduce sizable variance on the model performance. Indeed, the seed choice could influence weight initialization for task-specific layers when applying PLMs for downstream task [7]. When these task-specific layers are initialized poorly, coupled with small size of

---

[5] https://gluebenchmark.com/.

the training set, it is unlikely for the network to converge to a good solution without extensive training. In fact, previous work has shown that some seeds are consistently better than others for network optimization [7]. But it is not possible to know a priori which seed is better than the others.

3.2 Few-Shot Named Entity Recognition

After trying the fine-tuning methods above, we find that these methods still require hundreds of training examples, one question is whether it is possible to further reduce the number of training labels, so that only a few training labels are required. For example, the system designed by Yang and Katiyar [44], named StructShot, manages to beat competitive systems with a few samples, i.e. only 1 to 5 annotated named entities are required for each named entity class.

The key technique used in StructShot is the nearest neighbor search. Specifically, rather than directly training a PLM with token classification objective as we do, their system is first fine-tuned on a high-resource yet distantly related dataset; then they use the model as a feature extractor to obtain token representations of each of tokens in the test set and a small support set, both coming from low-resource task under investigation. With these two sets of representations in hand, test token's tag is determined through retrieving the most similar token's tag in the support set. A CRF decoding procedure follows the similarities estimated from previous steps to give the optimal tag sequence.

4 Experiments

4.1 Datasets

We experiment on the VIEW dataset provided by [8]. This dataset is collected from public vulnerability database Common Vulnerabilities and Exposure (CVE) and numerous security forums. The goal of this dataset is to identify vulnerable software names and software versions. Therefore, the tag set $\mathcal{Y}$ includes SN (software names) and SV (software versions). Similar to the other NER datasets, the O (outside) tag is used to represent all other tokens except vulnerable software names and versions. The dataset is manually labeled by three security experts in hope of minimizing labeling errors and enforcing consistent annotation standards. The records in dataset range in 20 years and contain all 13 categories listed on CVE website. The statistics of the dataset is shown in Table 2. We only report the average statistics of 12 categories here, and we refer readers interested in detailed statistics of these categories to Table 6. From the statistics of the dataset shown in Table 2, we observe the *dataset imbalance*, a challenge discussed in Sect. 2.3, in different levels.

- **Sample Number Imbalance** The memc category has significantly more samples in every split than the other 12 categories.

- **Entity Number Imbalance** The memc category has a predominately higher entity proportion in both sentence and token level than the other 12 categories.
- **Entity Type Imbalance** The appearances of SV tokens are generally more frequent than SN; sometimes this difference could vary up to 4.38% (see valid split of memc).

Table 2. Statistic of the VIEM dataset.

Category	memc			Average of other 12 categories		
Split	train	valid	test	train	valid	test
Number of sample	5758	1159	1001	468.00	116.67	556.25
Sentence-level entity proportion	0.5639	0.3287	0.4555	0.2435	0.2677	0.2620
Token-level entity proportion (SN)	0.0613	0.0368	0.0559	0.0214	0.0240	0.0237
Token-level entity proportion (SV)	0.0819	0.0807	0.0787	0.0308	0.0331	0.0331

Preparing for the Few-Sample Dataset. Among the 13 vulnerability categories in the dataset official release, only memc includes official validation split. In order to enable hyperparameter optimization and model selection on the other 12 categories, we randomly sample 10% of the training samples as the held-out validation set.

As we strive to investigate the performance of PLMs with the reduced number of annotated samples after fine-tuning and transfer learning, we create the training and validation set by sampling a subset of the full dataset. More specifically, for training set:

- **Fine-Tuning** We vary the proportion of sampled data from the memc training set from 1% to 10% (equivalent to a sample size from 58 to 576) and investigate the proportion to measure the adaptation of PLM to the computer security domain.
- **Transfer Learning** We vary the number of samples in the other 12 categories from $\{32, 64, 128, 256\}$ per category. Compared to fine-tuning, we retrain from sampling by proportion considering the fact that there are much smaller training set for the other 12 categories than the memc category.

For the validation set, we make sure it has the same size as the training set whenever possible: the infeasible cases are those when the size of the sampled training set exceeds the 10% limit. This comes from the consideration that the validation samples also require annotations and therefore consume labeling budget. Apart from the conversion of text form into data format required by model training, we do not perform any additional data processing to make sure the comparison is fair with previous system. Throughout the data preparation, we fix the random seed to 42.

4.2   Evaluation Metrics

Following the evaluation metrics used in [8], we use the precision, recall, and F1 score to evaluate our NER system, the definitions of precision, recall and F1 score are standard.

4.3   Experimental Setup

Hyperparameter Optimization Settings. We leverage the built-in API for hyperparameter optimization from the HugginFace library [42]. We leverage the grid search where the search space is as follows: the learning rate is selected from $\{1e-6, 5e-6, 1e-5\}$; the number of training epochs is chosen from $\{3, 5\}$; the batch size, is fixed to 2 per device because we find the resulting additional training iterations favorable as compared to setting it to 4 or 8 in our pilot experiment. All the other hyperparameters values are fixed to the default values in HuggingFace's `transformers-4.5.1`. Besides the grid search, we also experiment with the Bayesian optimization, but it fails to outperform the grid search with same computational resources. We fix the seed for model training to 42. We randomly restart every experiment 10 times and report the average score. We use the half-precision (i.e. `fp16`) mode to accelerate the training. During the hyperparameter optimization, we select the best checkpoint within each trial. For trials with different epoch numbers, we partition the training iterations so that the same number of model checkpoints are saved during training across different experiments. Due to the class imbalance problem (Sect. 4.1), measuring checkpoint quality solely based on metric of the SN or SV is suboptimal; therefore, we weigh their F1 score based on the number of their appearances in the groundtruth, i.e. N_{SN} and N_{SV}.

$$\bar{F}_1 = \frac{N_{\text{SN}} F_{\text{SN}} + N_{\text{SV}} F_{\text{SV}}}{N_{\text{SN}} + N_{\text{SV}}}$$

This single evaluation metric provides a tradeoff between metrics on different entities during model selection.

Evaluated Methods. Our experiments contain two parts: first, we evaluate the performance of few-sample learning on the `memc` category; second, we evaluate the performance of few-sample learning on the other 12 categories.

For the `memc` category, we fine-tune the models with varying proportions of the `memc` training data from 1% to 10%. We are interested to know whether fine-tuning with a small proportion of samples is feasible for PLMs to reach the performance of competitive system like `VIEM`.

For the other 12 categories, we compare the following 4 settings:

- **FT** This setting is the direct application of models fine-tuned on `memc` to the other categories. We expect the performance of FT to be the lowest across 4 settings.
- **FT+SS** This setting is consistent with the way this system is used in [44]: after the PLM is fine-tuned on the `memc` category, it is coupled with `StructShot` and directly applied to the other 12 categories without any transfer learning.
- **FT+TL** This setting is to apply transfer learning on other categories on the best fine-tuned checkpoint on the `memc` category. With the help of transfer learning, we expect the PLMs have satisfying performance across different vulnerability categories.
- **FT+TL+SS** This setting is built upon the **FT+TL** setting by allowing an additional application of `StructShot` on the model that experiences both fine-tuning and transfer learning. We expect a better performance of this setting over the **FT+TL** setting because of `StructShot`.

4.4   Experimental Results: Fine-Tuning on the memc Category

Summary of Main Results. The fine-tuning performance is detailed in Table 3. We could observe that the absolute difference of F1 score for both SN and SV are both within 1% absolute difference of the VIEM system. Importantly, the amount of data we use is only 10% of the VIEM system.

Table 3. Fine-tuning results for the memc category

	SN precision	SN recall	SN f1-score	SV precision	SV recall	SV f1-score
VIEM (full data)	0.9773	0.9916	0.9844	0.9880	0.9923	0.9901
BERT (10% data)	0.9582	0.9808	0.9694	0.9818	0.9848	0.9833
RoBERTa (10% data)	0.9567	0.9808	0.9686	0.9841	0.9844	0.9843
Electra (10% data)	0.9677	0.9754	0.9715	0.9830	0.9890	0.9860

To investigate how fine-tuning helps PLMs adapt to new domain and pick up task-awareness, we show a 2D projection of token-level hidden representation obtained on Electra for `memc` training set before and after fine-tuning in Fig. 2. We could see that much more compact clusterings of SN and SV are obtained; fine-tuning informs the PLMs with the underlying patterns of task dataset.

Fine-Tuning with Different Training Sample Size. We vary the number of samples per category from 1% to 10% as used in VIEM system to see how different sizes of training set affect the performance of fine-tuning.

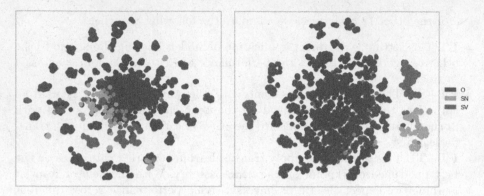

Fig. 2. The t-SNE visualization of `memc` training set in token level before (left) and after (right) fine-tuning. Fine-tuning gives more compact clusters for both SN and SV, illustrating the critical role of fine-tuning.

As is shown in Fig. 3, when the proportion of training samples increase from 1% to 10% for fine-tuning, the performance generally improves; but the improvement does not come monotonically. For example, in cases like fine-tuning RoBERTa with 5% of the data, the performance is worse than that of only 3%.

There is also a disparity on the fine-tuning performance of SN and SV. Specifically, the SV could have an F1 of around 95% with as small as 1% of the training samples available across three different models. This does not hold for the SN: under the 1% training data availability, the F1 for SN does not exceed 92% for BERT; for the RoBERTa and Electra, this metric is less than 88%. The disparity on the performance shows that correctly tagging SV is easier than tagging SN. This might arise from the fact that the there are naming conventions of software versions for the NER tagger to exploit while similar conventions do not exist for software names.

Another general tendency is the relation between precision and recall: across different models and the tag of interest, the recall is generally higher than precision. This is desirable during the deployment for security applications as incorrectly tagging regular software as vulnerabilities is acceptable, only leading to more manual efforts to double check individual software, while missing vulnerabilities could leave the critical systems unattended.

4.5   Experimental Results: Transfer Learning on the Other 12 Categories

Summary of the Main Results. As we could observe from the Table 4, transfer learning with the aggregation of 64 samples per category matches the performance of VIEM system; the absolute difference is less than 1% in F1 score. Despite less than 1% of improvement with respect to our system, the VIEM system is trained with the entire dataset with an aggregation of 7016 samples in total (see Fig. 6 for details; note the VIEM dataset merges train and valid split as its

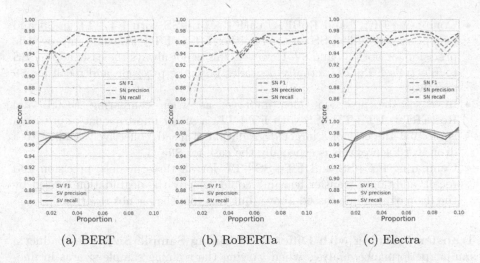

Fig. 3. The influence of training set size on fine-tuning.

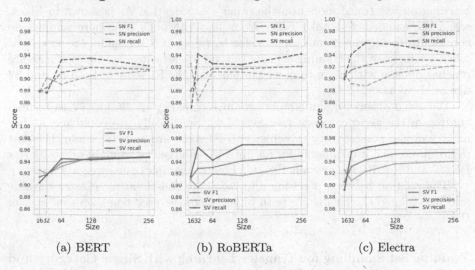

Fig. 4. The influence of training set size on transfer learning.

train split). Comparing the data utilization of our system to VIEM system, we reduce the number of annotations by 88.80%.

StructShot could help with the *precision* of SN but struggle with the *recall*. Introducing StuctShot has caused a degradation in F1 score, which casts the doubt on the applicability of this system to our application. Specifically, the absolute gap in F1 score could be up to 20% (compared to SN and SV's F1 scores in **FT+SS** and **FT+TL** for Electra). This gap still exists even with additional transfer learning (**FT+TL+SS**). Specifically,

- Comparing BERT and RoBERTa's **FT+TL+SS** with **FT+TL** in SN and SV's precision, adding **SS** on top of either **FT** or **FT+TL** could improve the precision for both SN and SV. However, this is not always the case: adding **SS** to Electra's **FT+TL** actually degrades both the precision and recall for SN and SV.
- In cases the precision does improve, as when comparing BERT and RoBERTa's **FT+TL+SS** with **FT+TL**, a minor improvement of precision is often at the cost of a larger degradation of recall, collectively leading to a drop in F1 score; this is especially the case for SN.
- Comparing **FT+SS** and **FT+SS+TL** for all three models' precision and recall, additional transfer learning could remedy the degradation caused by the drop of recall discussed above, but this tendency is not reverted.

Transfer Learning with Different Training Sample Size. We conduct a similar performance analysis when varying the training sample size as in fine-tuning; we visualize the average performance metrics since we apply transfer learning on fine-tuned model on the other 12 categories. As is shown in Fig. 4, when the number of samples range from 32, 64 through 256, more samples do contribute to improvements on transfer learning, but the gains are less clear when we increase the training set size beyond 64 per category: the performance reaches plateau with more than 64 samples (see the F1 score of both SN and SV across three models).

We could observe a similar performance disparity of SN and SV as in fine-tuning on the memc category (see Sect. 4.4). Specifically, we could use as few as 16 samples per category to achieve an F1 score around 91% while the same performance is not attainable for SN until we enlarge the training set to 64 (for BERT and RoBERTa) or even 128 (for Electra) samples per category.

The relation between precision and recall are consistent in transfer learning and fine-tuning: the recalls are both higher than the precision scores in most of the cases.

Training Set Sampling for Transfer Learning with Single Category and 12 Category Aggregate. The performance metrics detailed in Table 4 show that we could reach the performance of competitive systems by aggregating the 64 samples from each of the 12 categories, resulting in a training set In addition, we also evaluate the case where we only use the 64 examples in each category for training (and testing) without aggregation. We could see from the Table 5 that aggregating the training samples is favorable for transfer learning, improving the system performance by 2% to 3%.

Table 4. Transfer learning results averaged through 12 categories.

	SN precision	SN recall	SN f1-score	SV precision	SV recall	SV f1-score
VIEM (w/o transfer)	0.8278	0.8489	0.8382	0.8428	0.9407	0.8891
VIEM (w/ transfer)	0.9184	0.9286	0.9235	0.9382	0.9410	0.9396
BERT						
FT	0.8759	0.7950	0.8318	0.8813	0.9341	0.9060
FT+SS	0.9035	0.6843	0.7623	0.9001	0.7880	0.8183
FT+TL	0.8945	0.9302	0.9116	0.9373	0.9354	0.9355
FT+TL+SS	0.9060	0.7637	0.7766	0.9374	0.9135	0.9235
RoBERTa						
FT	0.8558	0.8510	0.8524	0.8822	0.9308	0.9052
FT+SS	0.8905	0.6810	0.7532	0.9106	0.8489	0.8735
FT+TL	0.8749	0.9409	0.9063	0.9162	0.9561	0.9354
FT+TL+SS	0.8841	0.7853	0.7989	0.9186	0.9364	0.9265
Electra						
FT	0.8656	0.8236	0.8416	0.8852	0.9246	0.9037
FT+SS	0.8692	0.6806	0.7274	0.8355	0.7428	0.7420
FT+TL	0.9003	0.9443	0.9214	0.9264	0.9494	0.9369
FT+TL+SS	0.8933	0.8263	0.8315	0.9035	0.9477	0.9233

Table 5. Comparison of transfer learning results with training set sampled from individual and aggregate of 12 categories. The results are averaged through 12 categories.

	SN precision	SN recall	SN f1-score	SV precision	SV recall	SV f1-score
RoBERTa						
FT+TL (individual)	0.8483	0.8953	0.8688	0.8816	0.9342	0.9059
FT+TL (aggregate)	0.8749	0.9409	0.9063	0.9162	0.9561	0.9354
Electra						
FT+TL (individual)	0.8670	0.9247	0.8940	0.9015	0.9346	0.9169
FT+TL (aggregate)	0.9003	0.9443	0.9214	0.9264	0.9494	0.9369

Analysis on the Two Data-Specific Challenges. We discussed in Sect. 4.1 that the VIEM dataset contains the challenges in *context dependency* and *dataset imbalance*. Even though we do not explicitly address either of them, our experiments show that they have not largely affected the performance. More specifically,

- **Dataset Imbalance** The average proportion of sentences with entities (SN and SV) in training set is only 24.35%; this might be a concern of performance due to the sparsity of informative tokens. However, as is shown in Table 4, the performance of models reaches or outperforms the VIEM system with an average F1 score of more than 0.9, which shows that the dataset imbalance problem does not hurt the effectiveness of our system.
- **Context Dependency** Our experiments show that transfer learning can help address the context dependency problem. Specifically, for the examples below, the model without fine-tuning misclassifies the tokens *Network Data*

Loss Prevention, Cisco Wireless LAN Controller, and *PineApp Mail-SeCure* as non-entities (O).

- CVE-2014-8523 (from `csrf` category): *Cross-site request forgery (CSRF) vulnerability in McAfee Network Data Loss Prevention (NDLP) before 9.3 allows remote ...*
- CVE-2015-0690 (from `xss` category): *Cross-site scripting (XSS) vulnerability ... on Cisco Wireless LAN Controller (WLC) devices ... aka Bug ID CSCun95178.*
- CVE-2013-4987 (from `gainpre` category): *PineApp Mail-SeCure before 3.70 allows remote authenticated ...*

Troubleshooting StructShot. The unsatisfactory performance of `StructShot` shown in Table 4 alerts caution to which scenario it could be applied. We therefore conduct an error analysis of `StructShot` to pinpoint its weakness. In our controlled experiment, we find that the `StructShot` is sensitive to adversarial examples. More specifically, as is shown in Fig. 5, when we manually create the support set by sequentially adding sample one after another, one single sample `0893580d85` could bring down overall performance by over 70%.

One benefit of the `StructShot` reported in [44] is its better capability of assigning non-entity tag (i.e. O tag) as these tokens do not carry unified semantic meaning. However, because of the *context dependency* illustrated in Fig. 1, in the NER of public security reports, non-entity tokens in one context might be entities in another one, causing confusion for the nearest neighbor-based tagger. We therefore suspect these confusing non-entities are the causes of the observed phenomenon. We monitor the following error cases to validate this hypothesis. Specifically, for CVE-2017-6038, the *Beldn Hirschmann GECKO Lite Managed switch* is originally tagged correctly as SN. However, when an additional `0893580d85` sample is introduced, all of its predictions are incorrectly reverted to O. Similarly for CVE-2014-9665, the correct predictions of software names *IB6131* and *Infiniband Switch* are changed to O due to the adversarial sample `0893580d85`.

- CVE-2017-6038: *A Cross-Site Request Forgery issue was discovered in Belden Hirschmann GECKO Lite Managed switch, Version 2.0.00 and prior versions.*
- CVE-2014-9565: *Cross-site request forgery (CSRF) vulnerability in IBM Flex System EN6131 40 Gb Ethernet and IB6131 40Gb Infiniband Switch 3.4.0000 and earlier.*

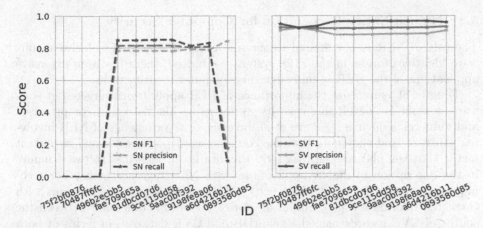

Fig. 5. `StructShot` is sensitive to adversarial samples. When testing on `csrf` category, an additional sample `0893580d85` in the support set could lead to a degradation of 70% of F1 score for `SN`.

5 Related Work

5.1 Information Extraction in Public Vulnerability Database

The `VIEM` system designed by Dong et al. is among the first batch of works that extracts information from public vulnerability database through combination of named entity recognition (NER) and relation extraction (RE) approaches. The extracted information is used to identify vulnerabilities and thereby inconsistencies between two major security databases - CVE and NVD [8].

Prior works of the information extraction in security domain mostly focus on other aspects like dataset curation, data analysis, and visualization. Specifically, Fan et al. create a C/C++ vulnerability dataset under the aid of information provided by the CVE database. Works by Ponta et al. and Wu et al. lead to similar datasets [9, 35]. All of them are facilitated by the CVE database during curation while the main differences are programming language and the dataset scale. Yamamoto et al. try to estimate the severity of vulnerability from the descriptions of CVE reports [43]. They also propose to improve the estimation accuracy by incorporating the relative importance of reports across different years. Pham and Dang implement an interactive tool named `CVExplorer` that enables the network and temporal view of different vulnerabilities in the CVE database [33].

This work follows the setting of `VIEM` system but tries to improve the vulnerability identification by minimizing the data labeling cost while matching the performance of existing systems.

5.2   Named Entity Recognition for Computer Security

The role of NER is not limited to serving as the first step of vulnerable software identification as in the VIEM system [8]. Indeed, there has been extensive applications of the NER in miscellaneous computer security tasks.

Feng et al. emphasize the importance of NER applications targeted at software domain for developing security applications like detecting compromised and vulnerable devices [11]. Ye et al. investigate the challenge of NER involving software, and they note that the common word polysemy is one important factor that complicates the task [46]. Pandita et al. identifies software names in the description of mobile applications with NER together with other techniques; then they locate the sentences pertaining to security permissions with extracted information [32]. Sun et al. notice the sizable difference between the source of CVE records named ExploitDB and CVE database in terms of both content completeness and update time. They, therefore, propose to automatically extract the relevant information from ExploitDB descriptions to create CVE records with NER and other technologies [38]. Ma et al. cast the API extraction from software documentation as an NER task, and they leverage an LSTM-based architecture to obtain high-fidelity APIs across different libraries and languages through fine-tuning and transfer learning [27]. Another work by Ye et al. attempts to use the NER to enable entity-centric applications for software engineering [45]. However, when they reduce their training data to 10% of the original dataset, their NER model's performance drop from an F1 score of 78% to 60%. This drop in performance highlights the difficulty of building NER systems in the few-shot setting.

Our work strives to identify vulnerabilities from public security reports, and therefore treats NER as a vehicle; furthermore, the architecture we leverage also benefits from fine-tuning and transfer learning. However, we face a unique challenge of resource constraints: our system should reach competitive systems' performance with as few labeled samples as possible.

5.3   Few-Sample Named Entity Recognition

The study of named entity recognition dates back to Message Understanding Conference-6 (MUC-6) held in 1995, where researchers first identified the problem of recognizing named entities such as names, organizations, locations using rule-based or probabilistic approaches [19]. But the adoption of neural approaches and the notion of few-sample NER only happened in recent years.

Lample et al. first propose to utilize neural networks in named entity recognition task and outperforms traditional CRF-based sequence labeling models [24]. Driven by their success, neural network has become the de facto choice for named entity recognition task since then. However, at the same time, the tension between cost of obtaining named entity annotations and data hunger nature of neural model training has gained attention. For example, Buys et al. notice that a graphical model like HMM could outperform its neural counterparts by more than 10% of accuracy in sequence tagging task given the access to same amount of training data [4]. Smaller number of parameters in HMM makes the training easier to converge than a RNN.

In order to tackle this challenge, the notion of few-shot learning, originally proposed by computer vision community [37, 40], gets noticed by NLP practitioners. These methods focus on inferring incoming samples' class membership based on prototypes obtained through metric learning. Their success in image classification propels the same methods to be adapted and tested in the NER task. Fritzler et al. are among the first to apply prototypical networks to NER task. However, their model treats each token independently by inferring one entity type at a time, disregarding the context of sentence [14]. Hou et al. extend this work and propose to merge specific named entity types into abstract classes (referred to as "collapsed transitions") to enable domain transfer [20]. Yang and Katiyar borrow the idea of collapsed transitions but simplify the overall architecture [44]. Rather than using dedicated architecture for few-shot learning, they first try to measure distance between tokens based on tokens' contextualized embeddings returned by PLM and then infer token's entity type based solely on nearest neighbor's types. After this step, an optional CRF decoding module is used to account for context of tokens. Their simple model architecture gain state-of-the-art performance compared with previous attempts. More importantly, besides the significant boost in performance metrics, their work shows the potential of pretrained language models (PLM) in reducing annotations by leveraging universal representations of language.

Another line of works consider the alternatives to the prototype-based approaches. Huang et al. show that self-training and active learning could help reduce data annotations [22]. Fries et al. provide empirical evidence of using data programming paradigm to convert rules given by the experts to named entity tags [13]. Li et al. propose a novel training protocol to make model-agnostic meta learning (MAML) framework, previously only available for sentence-level classification [1], applicable to sequence labeling (token-level classification) problem [25, 26].

Our work is consistent with previous works in the goal of reducing the number of training labels required. However, different from [44], we discard the tag assignment scheme based on prototypes and propose the direct application of PLMs through fine-tuning and transfer learning. The empirical evidence shows that this design choice attains better and stable performance metrics.

A distinct application scenario of few-sample learning is to test the system by inferring classes not seen in training time. We note that, despite both striving to generalize well to unseen samples during testing, our setting of reducing the required training examples is orthogonal to this goal following the taxonomy provided in [47].

6   Conclusions and Future Work

In this work, we take a first attempt to investigate the problem of few-sample named entity recognition (NER) for public security vulnerability reports. By leveraging fine-tuning of three pre-trained language models (BERT, RoBERTa, Electra), we find that it is possible to match the performance of prior art with much less labeled training data: first, for vulnerability reports in the memc category, fine-tuning of PLM can match the score of prior art by using only 10% of their training examples (or 576 labelled sentences); second, for vulnerability reports in the other 12 categories (e.g., bypass), we find that through fine-tuning and transfer learning, it is possible to match the scores of [8] by using only 11.2% of their training examples (or an aggregation of 64 labels from each category). As a result, few-sample learning has effectively reduced the training data required for NER for security vulnerability reports.

It is important to notice that our work is just a beginning and there exists great potential for further improvement. We identify future work in the following directions. First, leveraging unlabelled data. This work only considers labelled samples of named entity tags. It is worth exploring how to leverage unlabelled data to further improve the performance. It was shown in [29] that few-shot learning can achieve competitive performance on the GLUE dataset by leveraging models trained on a small amount of labelled data and predictions on a large amount of unlabelled examples as augmented noisy labels. As language models are usually pre-trained in the general domains, it can be expected that learning from unlabelled data in the security domain can help the model more quickly adapt to the in-domain task. Second, leveraging external knowledge base. Our experiment shows that it is more difficult to achieve a good F1 score for software name than software version. This result also meets our expectation, because there is a higher chance for a software name than software version in the testing dataset to be completely unseen. However, the language model may fail to bridge such gap since it is pre-trained on the general domain. To make the language model quickly recognize unseen software names, one approach is to leverage an external knowledge base on top of the few-sample learning model. It is an open question how to leverage the knowledge base to help with the few-sample learning without introducing many mismatched predictions. Third, by empirically observing the failure cases in transfer learning, we find that there exists some adversarial cases in the 12 sub-datasets (Sect. 4.5), which results in a dramatic drop in the performance of few-sample learning. One direction of future work is thus to investigate adversarial learning for the few-shot transfer learning to improve its robustness.

A   Dataset Statistics

Table 6 is the detailed version of Table 2. The valid split for categories other than memc is 10% sample of official train set. "Sentence-level entity proportion" refers to the proportion of sentences that have SN or SV, and "Token-level entity proportion" is proportion of SN and SV with respect to given dataset split. These proportions reflect the dataset imbalance in different levels.

Table 6. Detailed statistics of VIEM dataset.

Category	Split	Number of sample	Sentence-level entity proportion	Token-level entity proportion (SN)	Token-level entity proportion (SV)
memc	train	5758	0.5639	0.0613	0.0819
	valid	1159	0.3287	0.0368	0.0807
	test	1001	0.4555	0.0559	0.0787
bypass	train	652	0.2239	0.0314	0.0431
	valid	162	0.2469	0.0367	0.0423
	test	610	0.2902	0.0456	0.0531
csrf	train	521	0.2399	0.0207	0.0347
	valid	130	0.2846	0.0251	0.0397
	test	415	0.3181	0.0321	0.0464
dirtra	train	619	0.2359	0.0172	0.0219
	valid	155	0.1871	0.0180	0.0316
	test	646	0.2879	0.0197	0.0220
dos	train	396	0.2273	0.0212	0.0405
	valid	99	0.2020	0.0234	0.0419
	test	484	0.2624	0.0189	0.0331
execution	train	413	0.2639	0.0228	0.0358
	valid	103	0.2718	0.0314	0.0302
	test	639	0.2598	0.0273	0.0357
fileinc	train	546	0.2857	0.0175	0.0185
	valid	137	0.3869	0.0259	0.0222
	test	683	0.3133	0.0206	0.0215
gainpre	train	323	0.2229	0.0243	0.0430
	valid	80	0.3250	0.0357	0.0723
	test	577	0.2114	0.0191	0.0311
httprs	train	550	0.1891	0.0127	0.0217
	valid	137	0.1241	0.0077	0.0124
	test	411	0.2360	0.0175	0.0304
infor	train	305	0.2459	0.0326	0.0354
	valid	76	0.3158	0.0187	0.0282
	test	509	0.2358	0.0227	0.0348
overflow	train	396	0.2475	0.0217	0.0326
	valid	98	0.2143	0.0185	0.0230
	test	454	0.2819	0.0216	0.0343
sqli	train	538	0.2565	0.0145	0.0141
	valid	134	0.2836	0.0194	0.0151
	test	685	0.2423	0.0171	0.0181
xss	train	357	0.2829	0.0203	0.0289
	valid	89	0.3708	0.0276	0.0386
	test	562	0.2046	0.0219	0.0363

References

1. Bao, Y., Wu, M., Chang, S., Barzilay, R.: Few-shot text classification with distributional signatures, August 2019
2. Bridges, R.A., Jones, C.L., Iannacone, M.D., Testa, K.M., Goodall, J.R.: Automatic labeling for entity extraction in cyber security. arXiv preprint arXiv:1308.4941 (2013)
3. Brown, T.B., et al.: Language models are few-shot learners. arXiv preprint arXiv:2005.14165 (2020)

4. Buys, J., Bisk, Y., Choi, Y.: Bridging HMMs and RNNs through architectural transformations (2018)
5. Chang, J.R., Chen, Y.S.: Pyramid stereo matching network. In: Proceedings of the IEEE Conference on Computer Vision and Pattern Recognition, pp. 5410–5418 (2018)
6. Chen, H., Xia, M., Chen, D.: Non-parametric few-shot learning for word sense disambiguation. arXiv preprint arXiv:2104.12677 (2021)
7. Dodge, J., Ilharco, G., Schwartz, R., Farhadi, A., Hajishirzi, H., Smith, N.A.: Fine-tuning pretrained language models: weight initializations, data orders, and early stopping. arXiv abs/2002.06305 (2020)
8. Dong, Y., Guo, W., Chen, Y., Xing, X., Zhang, Y., Wang, G.: Towards the detection of inconsistencies in public security vulnerability reports. In: USENIX Security Symposium (2019)
9. Fan, J., Li, Y., Wang, S., Nguyen, T.N.: A C/C++ code vulnerability dataset with code changes and CVE summaries. In: Proceedings of the 17th International Conference on Mining Software Repositories, pp. 508–512. ACM, Seoul Republic of Korea, June 2020. https://doi.org/10.1145/3379597.3387501, https://dl.acm.org/doi/10.1145/3379597.3387501
10. Farhang, S., Kirdan, M.B., Laszka, A., Grossklags, J.: An empirical study of android security bulletins in different vendors. In: Proceedings of The Web Conference 2020, pp. 3063–3069 (2020)
11. Feng, X., Li, Q., Wang, H., Sun, L.: Acquisitional rule-based engine for discovering internet-of-thing devices. In: USENIX Security Symposium (2018)
12. Finn, C., Abbeel, P., Levine, S.: Model-agnostic meta-learning for fast adaptation of deep networks. In: International Conference on Machine Learning, pp. 1126–1135. PMLR (2017)
13. Fries, J., Wu, S., Ratner, A., Ré, C.: Swellshark: a generative model for biomedical named entity recognition without labeled data (2017)
14. Fritzler, A., Logacheva, V., Kretov, M.: Few-shot classification in named entity recognition task. In: Proceedings of the 34th ACM/SIGAPP Symposium on Applied Computing (2019)
15. Gaglione, G.S., Jr.: The equifax data breach: an opportunity to improve consumer protection and cybersecurity efforts in America. Buff. L. Rev. **67**, 1133 (2019)
16. Gao, P., et al.: A system for automated open-source threat intelligence gathering and management. arXiv preprint arXiv:2101.07769 (2021)
17. Gao, T., Fisch, A., Chen, D.: Making pre-trained language models better few-shot learners, December 2020
18. Gasmi, H., Laval, J., Bouras, A.: Information extraction of cybersecurity concepts: an LSTM approach. Appl. Sci. **9**(19), 3945 (2019)
19. Grishman, R., Sundheim, B.: Message understanding conference- 6: a brief history. In: COLING (1996)
20. Hou, Y., et al.: Few-shot slot tagging with collapsed dependency transfer and label-enhanced task-adaptive projection network. In: ACL (2020)
21. Hou, Y., et al.: Few-shot slot tagging with collapsed dependency transfer and label-enhanced task-adaptive projection network. arXiv preprint arXiv:2006.05702 (2020)
22. Huang, J., et al.: Few-shot named entity recognition: a comprehensive study. arXiv abs/2012.14978 (2020)
23. Johnson, J.M., Khoshgoftaar, T.M.: Survey on deep learning with class imbalance. J. Big Data **6**(1), 1–54 (2019). https://doi.org/10.1186/s40537-019-0192-5

24. Lample, G., Ballesteros, M., Subramanian, S., Kawakami, K., Dyer, C.: Neural architectures for named entity recognition (2016)
25. Li, J., Chiu, B., Feng, S., Wang, H.: Few-shot named entity recognition via meta-learning. IEEE Trans. Knowl. Data Eng. (2020). https://doi.org/10.1109/tkde.2020.3038670
26. Li, J., Shang, S., Shao, L.: MetaNER: named entity recognition with meta-learning. In: Proceedings of The Web Conference 2020. ACM, April 2020. https://doi.org/10.1145/3366423.3380127
27. Ma, S., Xing, Z., Chen, C., Qu, L., Li, G.: Easy-to-deploy API extraction by multi-level feature embedding and transfer learning. IEEE Trans. Softw. Eng. (2019). https://ieeexplore.ieee.org/document/8865646
28. Mosbach, M., Andriushchenko, M., Klakow, D.: On the stability of fine-tuning BERT: misconceptions, explanations, and strong baselines, June 2020
29. Mukherjee, S.S., Awadallah, A.H.: Uncertainty-aware self-training for few-shot text classification. In: NeurIPS 2020 (Spotlight). ACM, December 2020
30. Mulwad, V., Li, W., Joshi, A., Finin, T., Viswanathan, K.: Extracting information about security vulnerabilities from web text. In: 2011 IEEE/WIC/ACM International Conferences on Web Intelligence and Intelligent Agent Technology, vol. 3, pp. 257–260. IEEE (2011)
31. Palmer, C.: The Boeing 737 Max saga: automating failure. Engineering 6(1), 2–3 (2020). https://doi.org/10.1016/j.eng.2019.11.002
32. Pandita, R., Xiao, X., Yang, W., Enck, W., Xie, T.: WHYPER: towards automating risk assessment of mobile applications. In: USENIX Security Symposium (2013)
33. Pham, V., Dang, T.: CVExplorer: multidimensional visualization for common vulnerabilities and exposures. In: 2018 IEEE International Conference on Big Data (Big Data), pp. 1296–1301, December 2018. https://doi.org/10.1109/BigData.2018.8622092
34. Phang, J., Févry, T., Bowman, S.R.: Sentence encoders on stilts: supplementary training on intermediate labeled-data tasks. arXiv preprint arXiv:1811.01088 (2018)
35. Ponta, S.E., Plate, H., Sabetta, A., Bezzi, M., Dangremont, C.: A manually-curated dataset of fixes to vulnerabilities of open-source software. In: 2019 IEEE/ACM 16th International Conference on Mining Software Repositories (MSR), Montreal, QC, Canada, pp. 383–387. IEEE, May 2019. https://doi.org/10.1109/MSR.2019.00064, https://ieeexplore.ieee.org/document/8816802/
36. Snell, J., Swersky, K., Zemel, R.S.: Prototypical networks for few-shot learning. arXiv preprint arXiv:1703.05175 (2017)
37. Snell, J., Swersky, K., Zemel, R.S.: Prototypical networks for few-shot learning, March 2017
38. Sun, J., et al.: Generating informative CVE description from ExploitDB posts by extractive summarization. arXiv preprint arXiv:2101.01431 (2021)
39. Tjong Kim Sang, E.F., De Meulder, F.: Introduction to the CoNLL-2003 shared task: language-independent named entity recognition. In: Proceedings of the Seventh Conference on Natural Language Learning at HLT-NAACL 2003 - Volume 4. CONLL 2003, pp. 142–147. Association for Computational Linguistics, USA (2003). https://doi.org/10.3115/1119176.1119195
40. Vinyals, O., Blundell, C., Lillicrap, T., Kavukcuoglu, K., Wierstra, D.: Matching networks for one shot learning, June 2016
41. Wang, Y., Yao, Q., Kwok, J., Ni, L.: Generalizing from a few examples: a survey on few-shot learning. arXiv: Learning (2019)

42. Wolf, T., et al.: HuggingFace's transformers: state-of-the-art natural language processing. arXiv abs/1910.03771 (2019)
43. Yamamoto, Y., Miyamoto, D., Nakayama, M.: Text-mining approach for estimating vulnerability score. In: 2015 4th International Workshop on Building Analysis Datasets and Gathering Experience Returns for Security (BADGERS), Kyoto, Japan, pp. 67–73. IEEE, November 2015. https://doi.org/10.1109/BADGERS.2015.018, http://ieeexplore.ieee.org/document/7809535/
44. Yang, Y., Katiyar, A.: Simple and effective few-shot named entity recognition with structured nearest neighbor learning. arXiv abs/2010.02405 (2020)
45. Ye, D., Xing, Z., Foo, C.Y., Ang, Z.Q., Li, J., Kapre, N.: Software-specific named entity recognition in software engineering social content. In: 2016 IEEE 23rd International Conference on Software Analysis, Evolution, and Reengineering (SANER), vol. 1, pp. 90–101 (2016)
46. Ye, D., Xing, Z., Foo, C.Y., Li, J., Kapre, N.: Learning to extract API mentions from informal natural language discussions. In: 2016 IEEE International Conference on Software Maintenance and Evolution (ICSME), pp. 389–399 (2016)
47. Zhang, T., Wu, F., Katiyar, A., Weinberger, K.Q., Artzi, Y.: Revisiting few-sample BERT fine-tuning, June 2020
48. Zhou, C., Li, B., Sun, X.: Improving software bug-specific named entity recognition with deep neural network. J. Syst. Softw. **165**, 110572 (2020)

Malware Attack and Defense

DexRay: A Simple, yet Effective Deep Learning Approach to Android Malware Detection Based on Image Representation of Bytecode

Nadia Daoudi[✉], Jordan Samhi, Abdoul Kader Kabore, Kevin Allix, Tegawendé F. Bissyandé, and Jacques Klein

SnT, University of Luxembourg, 29, Avenue J.F Kennedy, 1359 Luxembourg, Luxembourg
{nadia.daoudi,jordan.samhi,abdoulkader.kabore, kevin.allix,tegawende.bissyande,jacques.klein}@uni.lu

Abstract. Computer vision has witnessed several advances in recent years, with unprecedented performance provided by deep representation learning research. Image formats thus appear attractive to other fields such as malware detection, where deep learning on images alleviates the need for comprehensively hand-crafted features generalising to different malware variants. We postulate that this research direction could become the next frontier in Android malware detection, and therefore requires a clear roadmap to ensure that new approaches indeed bring novel contributions. We contribute with a first building block by developing and assessing a baseline pipeline for image-based malware detection with straightforward steps.

We propose DexRay, which converts the bytecode of the app DEX files into grey-scale "vector" images and feeds them to a 1-dimensional Convolutional Neural Network model. We view DexRay as foundational due to the exceedingly basic nature of the design choices, allowing to infer what could be a minimal performance that can be obtained with image-based learning in malware detection.

The performance of DexRay evaluated on over 158k apps demonstrates that, while simple, our approach is effective with a high detection rate (F1-score = 0.96). Finally, we investigate the impact of time decay and image-resizing on the performance of DexRay and assess its resilience to obfuscation.

This **work-in-progress paper** contributes to the domain of Deep Learning based Malware detection by providing a sound, simple, yet effective approach (with available artefacts) that can be the basis to scope the many profound questions that will need to be investigated to fully develop this domain.

Keywords: Android security · Malware detection · Deep learning

© Springer Nature Switzerland AG 2021
G. Wang et al. (Eds.): MLHat 2021, CCIS 1482, pp. 81–106, 2021.
https://doi.org/10.1007/978-3-030-87839-9_4

1   Introduction

Automating malware detection is a key concern in the research and practice communities. There is indeed a huge number of samples to assess, making it challenging to consider any manual solutions. Consequently, several approaches have been proposed in the literature to automatically detect malware [1–4]. However, current approaches remain limited and detecting all malware is still considered an unattainable dream. A recent report from McAfee [5] shows that mobile malware has increased by 15% between the first and the second quarter of 2020. Moreover, Antivirus companies and Google, the official Android market maintainer, have disclosed that malware apps become more and more sophisticated and threaten users' privacy and security [6–8].

To prevent the spread of malware and help security analysts, researchers have leveraged Machine Learning approaches [9–15] that rely on features extracted statically, dynamically, or in an hybrid manner. While a dynamic approach extracts the features when the apps are running, a static approach relies exclusively on the artefacts present in the APK file. As for hybrid approaches, they combine both statically and dynamically retrieved features. Both static and dynamic approaches require manual engineering of the features in order to select some information that can, to some extent, approximate the behaviour of the apps. Also, the hand-crafted features might miss some information that is relevant to distinguish malware from benign apps. Unfortunately, good feature engineering remains an open problem in the literature due to the challenging task of characterising maliciousness in terms of app artefacts.

Recently, a new wave of research around representation of programs/software for malware detection has been triggered. Microsoft and Intel Labs have proposed STAMINA[1], a Deep Learning approach that detects malware based on image representation of binaries. This approach is built on deep transfer learning from computer vision, and it was shown to be highly effective in detecting malware. In the Android research community, image-based malware detection seems to be attractive. Some approaches have investigated the image representation of the APK's code along with deep learning techniques. However, they have directly jumped to non-trivial representations (e.g., colour) and/or leveraged complex learning architectures (e.g., Inception-v3 [16]). Furthermore, some supplementary processing steps are applied which may have some effects on the performance yielded. Besides, they create square or rectangular images, which might distort the bytecode sequences from the DEX files, therefore at the same time losing existing *locality* and creating artificial *locality*. Indeed, given that the succession of pixels in the image depends on the series of bytes in the DEX files, converting the bytecode to a "rectangular" image can result in having patterns that start at the end of a line (i.e., row of the image) and finish at the beginning of the next line in the image. Moreover, related approaches leverage

---

[1] https://www.microsoft.com/security/blog/2020/05/08/microsoft-researchers-work-with-intel-labs-to-explore-new-deep-learning-approaches-for-malware-classification/.

2-dimensional convolutional neural networks, which perform convolution operations that take into consideration the pixels in the 2-d neighbourhood using 2-d kernels. Since there is no relationship between pixels belonging to the same column (i.e., pixels that are above/below each others) of a "rectangular" image representation of code, the use of 2-d convolutional neural networks seems to be inappropriate.

Image-based representation is, still, a sweet spot for leveraging advances in deep learning as it has been demonstrated with the impressive results reported in the computer vision field. To ensure the development of the research direction within the community, we propose to investigate a straightforward "vector" image-based malware detection approach leveraging both a simple image representation and a basic neural network architecture. Our aim is to deliver a foundational framework towards enabling the community to build novel state-of-the-art approaches in malware detection. This paper presents the initial insights into the performance of a baseline that is made publicly available to the community.

Our paper makes the following contributions:

- We propose DEXRAY, a foundational image-based Android malware detection approach that converts the sequence of raw byte code contained in the DEX file(s) into a simple grey-scale "vector" image of size (1, 128*128). Features extraction and classification are assigned to a 1-dimensional-based convolutional neural network model.
- We evaluate DEXRAY on a large dataset of more than 158k malware and goodware apps.
- We discuss the performance achievement of DEXRAY in comparison with prior work, notably the state-of-the-art DREBIN approach as well as related work leveraging APK to image representations.
- We evaluate the performance of DEXRAY on detecting new malware apps.
- We investigate the impact of image-resizing on the performance of DEXRAY.
- We assess the resilience of DEXRAY and DREBIN to apps' obfuscation.
- We make the source code of DEXRAY publicly available along with the dataset of apps and images.

2 Approach

In this section, we present the main basic blocks of DEXRAY that are illustrated in Fig. 1. We present in Sect. 2.1 our image representation process which covers step (1.1) and (1.2) in Fig. 1. As for step (2), it is detailed in Sect. 2.2.

2.1 Image Representation of Android Apps

Android apps are delivered in the form of packages called APK (Android Package) whose size can easily reach several MB [17]. The APK includes the bytecode (DEX files), some resource files, native libraries, and other assets.

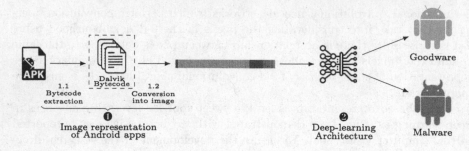

Fig. 1. DEXRAY basic building blocks.

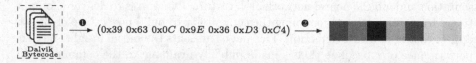

Fig. 2. Process of image generation from dalvik bytecode. ❶: bytecode bytes' vectorisation; ❷: Mapping bytes to pixels.

(1.1) Bytecode extraction: Our approach focuses on code, notably the applications' bytecode, i.e., DEX files, where the app behaviour is supposed to be implemented.

(1.2) Conversion into image: Our straightforward process for converting DEX files into images is presented in Fig. 2. We concatenate all the DEX files present in the APK as a single byte stream vector (step ❶ in Fig. 2). This vector is then converted to a grey-scale "vector" image by considering each byte as an 8-bit pixel (step ❷ in Fig. 2). Given that apps can widely differ in their code size, the size of the resulting "vector" images is also different. Note that the size of our "vector" image representation refers to the width of the image, since its height is 1. To comply with the constraints [18] of off-the-shelf deep learning architectures for images, we leverage a standard image resizing algorithm[2] to resize our "vector" images to a fixed-size. For our experiments, we have selected a size of $(1, 128 \times 128)$. We also investigate the impact of resizing on the performance of DEXRAY in Sect. 4.3.

We remind that related image-based Android malware detectors leverage a "rectangular" image representation, which might destroy the succession of bytes in the DEX files (i.e., the succession of pixels in the rectangular image). Moreover, this representation usually requires padding to have a rectangular form of the image. The complete procedure that shows the difference between our "vector" image and related approaches "rectangular" image generation is detailed in Algorithm 1 and Algorithm 2 respectively. While there is more advanced "rectangular" image representations, Algorithm 2 shows an illustration with a basic "square" grey-scale image of size $(128, 128)$.

---

[2] https://www.tensorflow.org/api_docs/python/tf/image/resize

---

Algorithm 1: Algorithm describing 8-bit grey-scale "vector" image generation from an APK

Input: APK file
Output: 8-bit grey-scale "vector" image of size (1, 128x128)

bytestream ← ∅
for *dexFile in APK* **do**
| bytestream ← bytestream + dexFile.toByteStream()
end
l ← bytestream.length()
img ← generate8bitGreyScaleVectorImage(bytestream, l)
img.resize_to_size(height=1, width=128x128)
img.save()

Algorithm 2: Algorithm describing 8-bit grey-scale "square" image generation from an APK

Input: APK file
Output: 8-bit grey-scale "square" image of size (128, 128)

bytestream ← ∅
for *dexFile in APK* **do**
| bytestream ← bytestream + dexFile.toByteStream()
end
l ← bytestream.length()
sqrt ← $\lceil \sqrt{l} \rceil$
sq ← sqrt^2
while *bytestream.length() ≠ sq* **do**
| bytestream ← bytestream + "\x00" // padding with zeros
end
// At this point, bytestream is divided in *sqrt* part of length *sqrt*
// In other words, it is represented as a *sqrt* × *sqrt* matrix
img ← generate8bitGreyScaleSquareImage(bytestream, sqrt)
img.resize_to_size(height=128, width=128)
img.save()

Comparing the two Algorithms, we can notice that our image representation is very basic since it uses the raw bytecode "as it is", without any segmentation or padding needed to achieve a "square" or "rectangular" form of the image. In the remainder of this paper, we use "image" to refer to our "vector" image representation of code.

2.2 Deep Learning Architecture

Convolutional Neural Networks (CNN) are specific architectures for deep learning. They have been particularly successful for learning tasks involving images as inputs. CNNs learn representations by including three types of layers: Convolution, Pooling, and Fully connected layers [19]. We describe these notions in the following:

- The convolution layer is the primary building block of convolutional neural networks. This layer extracts and learns relevant features from the input images. Specifically, a convolutional layer defines a number of filters (matrices) that detect patterns in the image. Each filter is convolved across the input image by calculating a dot product between the filter and the input. The filters' parameters are updated and learned during the network's training with the backpropagation algorithm [20,21]. The convolution operation creates feature maps that pass first through an activation function, and then they are received as inputs by the next layer to perform further calculation.
- A pooling layer is generally used after a convolutional layer. The aim of this layer is to downsize the feature maps, so the number of parameters and the time of computation is reduced [22]. Max pooling and Average pooling are two commonly used methods to reduce the size of the feature maps received by a pooling layer in CNNs [23].
- In the Fully connected layer, each neuron is connected to all the neurons in the previous and the next layer. The feature maps from the previous convolution/pooling layer are flattened and passed to this layer in order to make the classification decision.

Among the variety of Deep-Learning architectures presented in the literature, CNNs constitute a strong basis for deep learning with images. We further propose to keep a minimal configuration of the presented architecture by implementing a convolutional neural network model that makes use of 1-dimensional convolutional layers. In this type of layers, the filter is convolved across one dimension only, which reduces the number of parameters and the time of the training. Also, 1-d convolutional layers are the best suited for image representation of code since the pixels represent the succession of bytecode bytes' from the apps. The use of 1-d convolution on our images can be thought of as sliding a convolution window over the sequences of bytes searching for patterns that can distinguish malware and benign apps.

We present in Fig. 3 our proposed architecture, which contains two 1-dimensional convolutional/pooling layers that represent the extraction units of our neural network. The feature maps generated by the second max-pooling layer are then flattened and passed to a dense layer that learns to discriminate malware from benign images. The second dense layer outputs the detection decision.

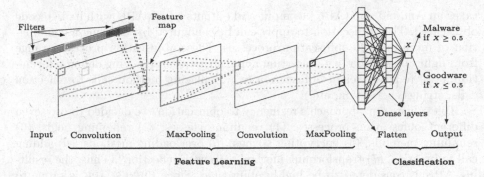

Fig. 3. Our Convolutional Neural Network architecture

3 Study Design

In this section, we first overview the research questions that we investigate. Then, we present the datasets and the experimental setup used to answer these research questions.

3.1 Research Questions

In this paper, we consider the following four main research questions:

- **RQ1:** How effective is DEXRAY in the detection of Android malware?
- **RQ2:** How effective is DEXRAY in detecting new Android malware?
- **RQ3:** What is the impact of image-resizing on the performance of DEXRAY?
- **RQ4:** How does app obfuscation affect the performance of DEXRAY?

3.2 Dataset

Initial Dataset. To conduct our experiments, we collect malware and benign apps from ANDROZOO [17] which contains, at the time of writing, more than 16 million Android apps. ANDROZOO crawls apps from several sources, including the official Android market Google Play[3]. We have collected ANDROZOO apps that have their compilation dates between January 2019 and May 2020. Specifically, our dataset contains 134 134 benign, and 71 194 malware apps. Benign apps are defined as the apps that have not been detected by any antivirus from VirusTotal[4]. The malware collection contains the apps that have been detected by at least two antivirus engines, similarly to what is done in DREBIN [9].

Obfuscation Process. In our experiments, we further explore the impact of app obfuscation. Therefore, we propose to generate obfuscated apps, by relying on the state-of-the-art Android app obfuscator OBFUSCAPK[5] [24]. This tool

---

[3] https://play.google.com/store.

[4] https://www.virustotal.com/.

[5] https://github.com/ClaudiuGeorgiu/Obfuscapk.

takes an Android app (APK) as input and outputs a new APK with its bytecode obfuscated. The obfuscation to apply can be configured by selecting a combination of more than 20 obfuscation processes. These obfuscation processes range from light obfuscation (e.g., inserting `nop` instructions, removing debug information, etc.) to heavyweight obfuscation (e.g., replacing method calls to reflection calls, strings encryption, etc.).

To evaluate our approach's resiliency to obfuscation, we decided to use seven different obfuscation processes: (1) renaming classes, (2) renaming fields, (3) renaming methods, (4) encrypting strings, (5) overloading methods, (6) adding call indirection, (7) transforming method calls with reflection[6]. Thus, the resulting APK is considered to be highly obfuscated. Since OBFUSCAPK is prone to crash, we were not able to get the obfuscated version for some apps.

We generate the images for the non-obfuscated dataset (the apps downloaded from ANDROZOO), and the obfuscated samples. The image generation process is detailed in Sect. 2.1.

Since we compare our method with DREBIN's approach, we have also extracted DREBIN's features from the exact same apps we use to evaluate our approach. In our experiments, we only consider the apps from the non-obfuscated and the obfuscated datasets for which we have (a) successfully generated their images, and (b) successfully extracted their features for DREBIN. Consequently, our final dataset contains 61 809 malware and 96 994 benign apps. We present in Table 1 a summary of our dataset.

Table 1. Dataset summary

	malicious apps	benign apps
Initial Set	71 194	134 134
Removed because of obfuscation failure	9023	34 629
Removed because of Image generation failure	4	2023
Removed because of DREBIN extraction failure	358	488
Final Set	61 809	96 994

3.3   Empirical Setup

Experimental Validation. We evaluate the performance of DEXRAY using the Hold-out technique [25]. Specifically, in all our experiments, we shuffle our dataset and split it into 80% training, 10% validation, and 10% test. We train our model on the training dataset, and we use the apps in the validation set to tune the hyper-parameters of the network. After the model is trained, we evaluate its performance using the apps in the test set. This process is repeated ten times by shuffling the dataset each time and splitting it into training, validation, and test sets. We repeat the Hold-out technique in order to verify that our results do not depend on a specific split of the dataset.

---

[6] https://www.oracle.com/technical-resources/articles/java/javareflection.html.

We set to 200 the maximum number of epochs to train the network, and we stop the training process if the accuracy score on the validation dataset does not improve for 50 consecutive epochs. As for the models' parameters, we use `kernel_size=12`, and `activation=relu` for the two convolution layers, and we set their number of filters to 64 and 128 respectively. We also use `pool_size=12` for the two max-pooling layers, and `activation=sigmoid` for the two dense layers. The number of neurons in the first dense layer is set to 64. As for the output layer, it contains one neuron that predicts the class of the apps. We rely on four performance measures to assess the effectiveness of DEXRAY: Accuracy, Precision, Recall, and F1-score. Our experiments are conducted using the TensorFlow library[7].

State-of-the-art Approaches to Compare with

DREBIN [9]: We compare DEXRAY against DREBIN, the state-of-the-art approach in machine learning based malware detection. DREBIN extracts features that belong to eight categories: Hardware components, Requested permissions, App components, Filtered intents, Restricted API calls, Used permissions, Suspicious API calls, and Network addresses. These features are extracted from the disassembled bytecode and the Manifest file. The combination of extracted features is used to create an n-dimensional vector space where n is the total number of extracted features. For each app in the dataset, an n-dimensional binary vector space is created. A value of 1 in the vector space indicates that the feature is present in the app. If a feature does not exist in the app, its value is set to 0. DREBIN feeds the vectors space to a Linear SVM classifier, and it uses the trained model to predict if an app is malware or goodware. In this study, we use a replicated version [26] of DREBIN that we run on our dataset.

R2-D2 [27] is an Android malware detector that converts the bytecode of an APK file (i.e., DEX files) into RGB colour images. Specifically, the hexadecimal from the bytecode is translated to RGB colour. R2-D2 is a CNN-based approach that trains Inception-v3 [16] with the coloured images to predict malware. In their paper, the authors state that their approach is trained with more than 1.5 million samples. However, they do not clearly explicit how many apps are used in the test nor the size fixed for the coloured images. The authors provide a link to the materials of their experiment[8], but we could not find the image generator nor the original apps used to evaluate their approach in Section IV-C of their paper. Only the generated images are available. As a result, we were unable to reproduce their experiment to compare with our model. Instead, we compare directly with the results reported in their paper.

Ding et al. [28] proposes to convert the DEX files into (512, 512) grey-scale images in order to feed them to a deep learning model. The authors experiment with two CNN-based models: The first one, we note `Model1`, contains four convolutional layers, four pooling layers, a fully-connected hidden layer, and a fully-connected output layer. `Model2`, which is the second model, has the same architecture as `Model1` but with an additional high-order feature layer that is

---

[7] https://www.tensorflow.org.
[8] http://R2D2.TWMAN.ORG.

added just after the first pooling layer. Basically, this layer contains the multiplication of each two adjacent feature maps from the previous layer, as well as the feature maps themselves. Since neither the dataset nor the implementation of Ding et al.'s models is publicly available, we also rely on the results reported in Ding et al.'s manuscript.

4   Study Results

4.1   RQ1: How Effective is DEXRAY in the Detection of Android Malware?

In this section, we assess the performance of DEXRAY on Android malware detection. We consider the performance against a ground truth dataset (the non-obfuscated apps introduced in Sect. 3.2) as well as a comparison against prior approaches. We rely on the experimental setup presented in Sect. 3.3 and we test the performance of DEXRAY on 15 880 apps (10% of the non-obfuscated apps).

We report in Table 2 the scores as the average of Accuracy, Precision, Recall, and F1-score.

Table 2. Performance of DEXRAY against DREBIN on our experimental dataset

	Accuracy	Precision	Recall	F1-score
DEXRAY	0.97	0.97	0.95	0.96
DREBIN	0.97	0.97	0.94	0.96

Overall, as shown in Table 2, DEXRAY reaches an average score of 0.97, 0.97, 0.95, and 0.96 for Accuracy, Precision, Recall, and F1-score respectively. The reported results show the high effectiveness of DEXRAY in detecting Android malware. We further compare the performance of DEXRAY against three Android malware detectors in the following.

Comparison with DREBIN. To assess the effectiveness of DEXRAY, we compare our results against a state-of-the-art Android malware detector that relies on static analysis: DREBIN. Specifically, we evaluate DREBIN using the same exact non-obfuscated apps we use to evaluate DEXRAY, and the same experimental setup described in Sect. 3.3. Moreover, we use the same split of the dataset for the Hold-out technique to evaluate the two approaches. We report the average of Accuracy, Precision, Recall, and F1-score of DREBIN's evaluation in Table 2.

We notice that DREBIN and DEXRAY achieve the same Accuracy, Precision, and the F1-score. As for the Recall DEXRAY slightly outperforms DREBIN with a difference of 0.01.

Table 3. Authors-reported performance of R2-D2 and Ding et al. on different and less-significant datasets

	Accuracy	Precision	Recall	F1-score
DEXRAY	0.97	0.97	0.95	0.96
R2-D2	0.97	0.96	0.97	0.97
Ding et al.-Model 1	0.94	–	0.93	–
Ding et al.-Model 2	0.95	–	0.94	–

Comparison Against Other Image-Based Malware Detection Approaches. We present in Table 3 the detection performance of R2-D2 and Ding et al.'s approaches as reported in their original publications. We note that DEXRAY and these two approaches are not evaluated using the same experimental setup and dataset. Specifically, R2-D2 is trained on a huge collection of 1.5 million apps, but it is evaluated on a small collection of 5482 images. In our experiments, we have conducted an evaluation on 15 880 test apps. We note that we have inferred the size of the test set based on R2-D2 publicly available images. Also, the scores we report for DEXRAY are the average of the scores achieved by ten different classifiers, each of which is evaluated on different test samples. R2-D2 scores are the results of a single train/test experiment which makes it difficult to properly compare the two approaches.

Similarly, Ding et al.'s experiments are conducted using the cross-validation technique, and both `Model1` and `Model2` are trained and evaluated using a small dataset of 4962 samples. Overall, we note that in terms of Recall, Precision and Accuracy, DEXRAY achieves performance metrics that are on par with prior work.

> **RQ1 answer:** DEXRAY is a straightforward approach to malware detection which yields performance metrics that are comparable to the state of the art. Furthermore, it demonstrates that its simplicity in image generation and network architecture has not hindered its performance when compared to similar works presenting sophisticated configurations

4.2 RQ2: How Effective is DEXRAY in Detecting New Android Malware?

Time decay [29] or model ageing [30] refers to the situation when the performance of ML classifiers drops after they are tested on new samples [31]. In this section, we aim to assess how does model ageing affect the performance of DEXRAY. Specifically, we investigate if DEXRAY can detect new malware when all the samples in its training set are older than the samples in its test set—a setting that Tesseract authors called *Temporally consistent* [29]. To this end, we split our dataset into two parts based on the date specified in the DEX file of the apps. The apps from 2019 are used to train and tune the hyper-parameters of the

Table 4. Impact of model ageing on the performance of DEXRAY

	Accuracy	Precision	Recall	F1-score
DEXRAY results from RQ1	0.97	0.97	0.95	0.96
DEXRAY (Temporally Consistent)	0.97	0.97	0.98	0.98

model, and the apps from 2020 are used for the test. The training and validation datasets consist of 113 951 and 28 488 apps respectively (i.e., 80% and 20% of 2019 dataset). As for the test dataset, it contains 16 364 malware and benign apps from 2020. We report our results on Table 4.

We notice that DEXRAY detects new malware apps with a high detection rate. Specifically, it achieves detection scores of 0.97, 0.97, 0.98, 0.98 for Accuracy, Precision, Recall, and F1-score respectively. Compared to its effectiveness reported in RQ1 in Sect. 4.1, we notice that DEXRAY has reported higher Recall in this Temporally consistent experiment. This result could be explained by the composition of the malware in the training and test set. Indeed, malicious patterns in the test apps have been learned during the training that contains a representative set of Android malware from January to December 2019. The high performance of DEXRAY on new Android apps demonstrates its ability to generalise and its robustness against model ageing.

In this experiment, it is not possible to use the 10-times Hold-out technique because we only have one model trained on apps from 2019 and tested on new apps from 2020. To conduct an in-depth evaluation of the detection capabilities of this model, we also examine the receiver operating characteristic (ROC) curve. The ROC curve shows the impact of varying the threshold of the positive class on the performance of the classifier. It can be generated by plotting the true positive rate (TPR) against the false positive rate (FPR). We present in Fig. 4 the ROC curve of our model.

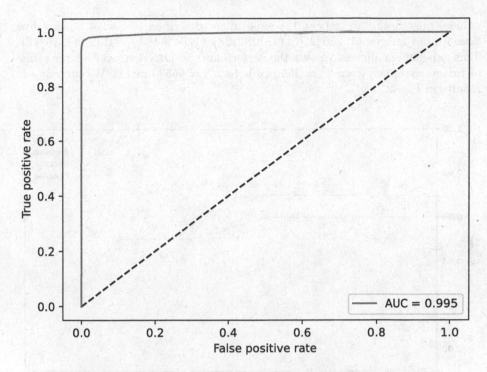

Fig. 4. ROC curve of DexRay evaluated against time decay

The ROC curve of DexRay further confirms its high effectiveness. As the area under the curve (AUC) can have a maximum value of 1, our approach has reached a very high AUC of 0.995.

It is noteworthy to remind that Tesseract [29] reported that DREBIN performance was very significantly (negatively) impacted when tested in a *Temporally Consistent* setting.

> **RQ2 answer:** DexRay's high performance is maintained in a *Temporally Consistent* evaluation. When tested on new Android apps, our approach has achieved very high detection performance, with an AUC of 0.995.

4.3   RQ3: What is the Impact of Image-Resizing on the Performance of DexRay?

In this section, we study the impact of image-resizing on the effectiveness of our approach. As we have presented in Sect. 2.1, we resort to resizing after mapping the bytecode bytes' to pixels in the image. Since the DEX files can have different sizes, resizing all images to the same size is necessary to feed our neural network. Resizing implies a loss of information. To better assess the impact of image-resizing, we evaluate the performance of DexRay using different image size values. Specifically, we repeat the experiment described in Sect. 4.1 using five sizes for our images: (1, 256*256), (1, 128*128), (1, 64*64), (1, 32*32), (1, 16*16). This experiment allows to assess the performance of DexRay over a large range of image sizes, i.e., from $2^8 = 256$ pixels to $2^{16} = 65536$ pixels. We present our results in Fig. 5.

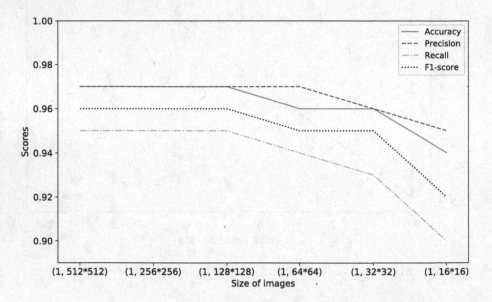

Fig. 5. The impact of image-resizing on the performance of DexRay

Overall, the size of images is a significant factor in the performance of DexRay. Unsurprisingly, the general trend is that the performance decreases as the image size is reduced: We notice that the values of the four metrics are lower for the three sizes that are smaller than our baseline (1, 128*128), with the worst performance being obtained with the smallest images. However, increasing the size from (1, 128*128) to either (1, 256*256) or (1, 512*512) does not improve the performance.

> **RQ3 answer:** Image-resizing has a significant impact on the effectiveness of our approach. While downsizing decreases the evaluation scores values by up to 5%, increasing the size of our images does not bring significant performance benefits. Hence (1, 128*128) seems to be the sweet spot of DexRay.

4.4   RQ4: How Does App Obfuscation Affect the Performance of DexRay?

Malware detectors in the literature are often challenged when presented with obfuscated apps. We propose to investigate to what extent DexRay is affected by obfuscation. We consider two scenarios where: (1) the test set includes obfuscated apps; (2) the training set is augmented with obfuscated samples.

Performance on Obfuscated Apps When DexRay **is Trained on a Dataset of Non-obfuscated Apps.** While our non-obfuscated dataset may contain obfuscated samples, we consider it as a normal dataset for training and we assess the performance of DexRay when the detection targets obfuscated apps. Specifically, we conduct two variant experiments to assess whether DexRay can detect obfuscated malware when: (1) It is tested on obfuscated apps that it has seen their non-obfuscated version in the training dataset, and (2) It is tested on obfuscated apps that it has NOT seen their non-obfuscated version in the training dataset. We consider both the non-obfuscated and the obfuscated samples, and we perform our experiments using the 10-times Hold-out technique described in Sect. 3.3.

For the first experiment, the non-obfuscated dataset is split into 90% training and 10% validation. The test set, which we note Test1, includes all the obfuscated apps. As for the second experiment, we design it as follows: We split the non-obfuscated dataset into 80% training, 10% validation, and 10% test. The training and the validation apps are used to train and tune DexRay hyperparameters. We do not rely on the test images themselves, but we consider their obfuscated versions, which we note Test2. The average scores of the two experiments are presented in Table 5.

We have also evaluated Drebin using the same experimental setup in order to compare with DexRay. Its results are also presented in Table 5.

Table 5. Performance of DexRay & Drebin on the obfuscated apps

	Accuracy	Precision	Recall	F1-score
DexRay evaluated on Test1	0.64	0.64	0.17	0.26
DexRay evaluated on Test2	0.64	0.65	0.13	0.22
Drebin evaluated on Test1	0.94	0.94	0.91	0.93
Drebin evaluated on Test2	0.94	0.93	0.90	0.92

We notice that in the two experiments, DexRay does not perform well on the obfuscated apps detection. Its scores reported previously in Table 2 for malware detection have dropped remarkably, especially for the Recall that is decreased by at least 0.78. The comparison of Drebin's detection scores in Table 5 and Table 2 suggests that its effectiveness is slightly decreased on the obfuscated apps detection. The scores reported in Table 5 are all above 0.9, which demonstrates that Drebin's overall performance on the obfuscated apps is still good. Drebin's results can be explained by the fact that it relies on some features that are not affected by the obfuscation process (e.g., requested permissions).

In the rest of this section, we investigate whether augmenting the training dataset with obfuscated samples can help DexRay discriminate the obfuscated malware.

Can Augmenting the Training Dataset with Obfuscated Samples Help to Discriminate Malware?

With this RQ, we aim to investigate if we can boost DEXRAY's detection via augmenting the training dataset with obfuscated apps. Specifically, we examine if this data augmentation can improve: (1) Obfuscated malware detection, and (2) Malware detection.

We conduct our experiments as follows: We split the non-obfuscated dataset into three subsets: 80% for the training, 10% for the validation, and 10% for the test. We augment the training and the validation subsets with X% of their obfuscated versions from the obfuscated dataset, with X = {25, 50, 75, 100}. As for the test dataset, we evaluate DEXRAY on both the non-obfuscated images and their obfuscated versions separately. Specifically, we assess whether augmenting the dataset with obfuscated apps can further enhance: (1) DEXRAY's performance on the detection of obfuscated malware reported in Table 5 (the test set is the obfuscated samples), and (2) DEXRAY's effectiveness reported in Table 2 (the test set contains non-obfuscated apps).

Similarly, we evaluate DREBIN on the same experimental setup, and we report the average prediction scores of both DEXRAY and DREBIN in Table 6.

Table 6. Performance of DEXRAY & DREBIN after dataset augmentation

		Accuracy	Precision	Recall	F1-score
DEXRAY tested on Obf-apps	25%	0.95	0.96	0.92	0.94
	50%	0.96	0.97	0.92	0.95
	75%	0.96	0.97	0.93	0.95
	100%	0.96	0.97	0.94	0.95
DREBIN tested on Obf-apps	25%	0.96	0.97	0.93	0.95
	50%	0.96	0.97	0.94	0.95
	75%	0.97	0.97	0.94	0.96
	100%	0.97	0.97	0.94	0.96
DEXRAY tested on non-Obf-apps	25%	0.97	0.97	0.94	0.96
	50%	0.97	0.97	0.94	0.96
	75%	0.97	0.97	0.94	0.96
	100%	0.97	0.97	0.94	0.95
DREBIN test on non-Obf-apps	25%	0.97	0.97	0.94	0.96
	50%	0.97	0.97	0.94	0.96
	75%	0.97	0.97	0.94	0.96
	100%	0.97	0.97	0.94	0.96

We can see that DEXRAY detection scores on obfuscated samples increase remarkably when adding obfuscated apps to the training dataset. With 100%

data augmentation, DEXRAY achieves a detection performance that is comparable to its performance on malware detection reported in Table 2. As for the
impact of data augmentation on malware detection, we notice that the detection
scores are stable. These results suggest that data augmentation boosts DEXRAY
detection on obfuscated malware, but it does not affect its performance on malware detection (non-obfuscated apps).

As for DREBIN, we notice that its performance on the obfuscated apps is
also improved, and it is comparable to its performance on malware detection
reported in Table 2. Similarly, its effectiveness on the non-Obfuscated apps after
data augmentation is not enhanced, but it is stable.

> **RQ4 answer:** Obfuscation can have a significant impact on the performance
> of DEXRAY. When the training set does not include obfuscated apps, the per-
> formance on obfuscated samples is significantly reduced. However, when the
> training set is (even slightly) augmented with obfuscated samples, DEXRAY
> can maintain its high detection rate.

5 Discussion

5.1 Simple But Effective

Prior work that propose image-based approaches to Android malware detection build on advanced "rectangular" image representations and/or complex network architectures. While they achieve high detection rates, the various levels of
sophistication in different steps may *hide* the intrinsic contributions of the basic
underlying ideas. With DEXRAY, we demonstrate that a minimal approach (i.e.,
with straightforward image generation and a basic CNN architecture) can produce high detection rates in detecting Android malware.

Our experimental comparison against the state-of-the-art detector DREBIN
further reveals that DEXRAY is competitive against DREBIN. In the absence
of artefacts to reproduce R2-D2 and Ding et al.'s approaches, our comparison focused on the detection scores reported by the authors. Note that while
DEXRAY yields similar scores, both approaches involve a certain level of complexity: they both rely on 2-dimensional convolution that needs more computational requirements than the simple form of convolution leveraged by DEXRAY.
Moreover, Ding et al.'s best architecture includes a high-order feature layer that
is added to four extraction units (convolution/pooling layers). As for R2-D2,
in addition to the coloured images that require a three-channel representation,
it already leverages a sophisticated convolutional neural network with 42 layers [16]. Besides, 2-d convolution may not be suitable for image representation
of code since pixels from one column of the image are not related. Such sophisticated design choices might affect the real capabilities of image-based malware
detectors.

5.2   The Next Frontier in Malware Detection?

Selecting the best features for malware detection is an open question. Generally, every new approach to malware detection comes with its set of features, which are a combination of some known features and some novel hand-crafted feature engineering process. Manual feature engineering is however challenging and the contributions of various features remain poorly studied in the literature.

Recent deep learning approaches promise to automate the generation of features for malware detection. Nevertheless, many existing approaches still perform some form of feature selection (e.g., APIs, AST, etc.). Image-based representations of code are therefore an appealing research direction in learning features without a priori selection.

DEXRAY's effectiveness suggests that deep learned feature sets could lead to detectors that outperform those created with hand-crafted features. With DEXRAY, we use only the information contained in the DEX file, but still we achieve a detection performance comparable to the state of the art in the literature. This research direction therefore presents a huge potential for further breakthroughs in Android malware detection. For instance, the detection capability of DEXRAY can be further boosted using the image representation of other files from the Android APKs (e.g., the Manifest file). We have also revealed that DEXRAY is not resilient to obfuscation, which calls for investigations into adapted image representations and neural network architectures. Nevertheless, we have demonstrated that the performance of DEXRAY is not affected by the time decay. Overall, emerging image-based deep learning approaches to malware detection are promising as the next frontier of research in the field: with the emergence of new variants of malware, automated deep feature learning can overcome prior challenges in the literature for anticipating the engineering of relevant features to curb the spread of malware.

5.3   Explainability and Location Concerns

Explainable AI is an increasingly important concern in the research community. In deep learning based malware detection, the lack of explainability may hamper adoption due to the lack of information to enable analysts to validate detectors' performance. Specifically, with DEXRAY, we wonder: how even a straightforward approach can detect malware with such effectiveness? Are malicious behaviours that easy to distinguish? What do image-based malware detectors actually learn? The results of our work call for further investigation of the explainability of such approaches. Since neural networks are black-box models, explanation methods have been proposed to interpret their predictions (e.g., LIME [32], LEMNA [33]). Leveraging these explanation methods for image-based malware detection can serve to interpret their prediction, and further advance this research direction.

Following up on the classical case of the wolf and husky detector that turned out to be a snow detector [32], we have concerns as to whether image-based malware detectors learn patterns that humans can interpret as signs of maliciousness. A more general question that is raised is whether such approaches can eventually help to identify the relevant code that implements the malicious behaviour in an app. Malware location is indeed important as it is essential for characterising variants and assessing the severity of maliciousness.

5.4   Threats to Validity

Our study and conclusions bear some threats to validity that we attempted to mitigate.

Threats to **external validity** refer to the generalisability of our findings. Our malware dataset indeed may not be representative of the population of malware in the wild. We minimised this threat by considering a large random sampling from AndroZoo, and by further requiring the consensus of only 2 antivirus engines to decide on maliciousness. Similarly, our results may have been biased by the proportion of obfuscated samples in our dataset. We have mitigated this threat by performing a study on the impact of obfuscation.

Threats to **internal validity** relate to the implementation issues in our approach. First, DEXRAY does not consider code outside of the DEX file. While this is common in current detection approaches that represent apps as images, it remains an important threat to validity due to the possibility to implement malware behaviour outside the main DEX files. Future studies should investigate apk to image representations that account for all artefacts. Second, we have relied on third-party tools (e.g., OBFUSCAPK), which fail on some apps, leading us to discard apps that we were not able to obfuscate or for which we cannot generate images. This threat is however mitigated by our selection of large dataset for experiments. Finally, setting the parameters of our experiments may create some threats to validity. For example, since Android apps differ in size, the generated images also have different sizes, which requires to resize all images in order to feed the Neural Network. The impact of image-resizing on the performance of our approach has been investigated in Sect. 4.3.

6   Related Work

Since the emergence of the first Android malware apps more than ten years ago [34], several researchers have dedicated their attention to develop approaches to stop the spread of malware. Machine Learning and Deep Learning techniques have been extensively leveraged by malware detectors using features extracted either using static, dynamic or hybrid analysis. We present these approaches in Sect. 6.1 and Sect. 6.2. We also review related work that leverages the image representation of code for malware detection in Sect. 6.3.

6.1   Machine Learning-Based Android Malware Detection

Recent studies have been proposed to review and summarise the research advancement in the field of machine learning-based Android malware detection [35, 36]. In 2014, DREBIN [9] has made a breakthrough in the field by proposing a custom feature set based on static analysis. DREBIN trains a Linear SVM classifier with features extracted from the DEX files and the Manifest file. Similarly, a large variety of static analysis-based features are engineered and fed to machine learning techniques to detect Android malware (e.g., MaMaDroid [11], RevealDroid [10], DroidMat [37], ANASTASIA [12]). Dynamic analysis has also been leveraged to design features for malware detection (e.g., DroidDolphin [14], Crowdroid [38], DroidCat [13]). While the above detectors rely either on static or dynamic analysis, some researchers have chosen to rely on features that combine the two techniques (e.g., Androtomist [39], BRIDEMAID [15], SAMADroid [40]).

All the referenced approaches require feature engineering and pre-processing steps that can significantly increase the complexity of the approach. Our method learns from raw data and extracts features automatically during the training process.

6.2 Deep Learning-Based Android Malware Detection

Deep Learning techniques have been largely adopted in the development of Android malware detectors. They are anticipated to detect emerging malware that might escape the detection of the conventional classifiers [41]. A recent review about Android malware detectors that rely on deep learning networks has been proposed [42]. MalDozer [43], a multimodal deep learning-based detector [44], DroidDetector [45], DL-Droid [46], Deep4MalDroid [47], and a deep autoencoder-based approach [48] are examples of DL-based detectors that predict malware using a variety of hand-crafted features. For instance, MalDozer [43] is a malware detection approach that uses as·features the sequences of API calls from the DEX file. Each API call is mapped to an identifier stored in a specific dictionary. MalDozer then trains an embedding word model word2vec [49] to generate the feature vectors. DroidDetector [45] is a deep learning-based Android malware detector that extracts features based on hybrid analysis. Required permissions, sensitive APIs, and dynamic behaviours are extracted and fed to the deep learning model that contains an unsupervised pre-training phase and a supervised back-propagation phases. Opcode sequences have been extracted and projected into an embedding space before they are fed to a CNN-based model [50]. Again, most of these approaches require a feature engineering step and/or huge computation requirement that is not needed by DEXRAY.

6.3 Image-Based Malware Detection

Different than traditional methods, some approaches have been proposed to detect malware based on the classical "rectangular" image-representation of source code. In 2011, a malware detection approach [51] that converts malware binaries into grey images, and extracts texture features using GIST [52] has been proposed. The features are fed to a KNN algorithm for classification purposes. Another study [53] has considered the same features for Android malware detection. Recently, a study [54] has been published about the use of image-based malware analysis with deep learning techniques.

In the Android ecosystem, an malware detection approach [55] with three types of images extracted from (1) Manifest file, (2) DEX file, and (3) Manifest, DEX and Resource.arsc files has been proposed. From each type of image, three global feature vectors and four local feature vectors are created. The global feature vectors are concatenated in one feature vector, and the bag of visual words algorithm is used to generate one local feature vector. The two types of vectors for the three types of images have been used to conduct a set of malware classification experiments with six machine learning algorithms. Similarly, an Android (and Apple) malware detector [56] trained using features extracted from grey-scale binary images has been proposed. The method creates the histogram of images based on the intensity of pixels and converts the histogram to 256 features vectors. The authors have experimented with different deep learning architectures, and the best results are achieved using a model with ten layers. R2D2 [27] and Ding et al. [28] have proposed malware detectors that are also based on image-learning and they are discussed in detail in Sect. 3.3. None of the above approaches has considered an image representation that preserve the continuity of the bytecode in DEX files. Moreover, they perform further pre-processing on the images before automatically extracting the features or rely on sophisticated ML or DL models for features extraction and classification. Our method converts the raw bytecode to "vector" images and feeds them directly to a simple 1-dimensional-based CNN model for features extraction and classification.

7   Conclusion

We have conducted an investigation to assess the feasibility of leveraging a simple and straightforward approach to malware detection based on image representation of code. DEXRAY implements a 1-dimensional convolution with two extraction units (Convolution/Pooling layers) for the neural network architecture. The evaluation of DEXRAY on a large dataset of malware and benign android apps demonstrates that it achieves a high detection rate. We have also showed that our approach is robust against time decay, and studied the impact of image-resizing on its performance. Moreover, we have investigated the impact of obfuscation on the effectiveness of DEXRAY and demonstrated that its performance can be further enhanced by augmenting the training dataset with obfuscated apps.

We have also compared DEXRAY against prior work on Android malware detection. Our results demonstrate that DEXRAY performs similarly to the state of the art DREBIN and two image-based detectors that consider more sophisticated network architectures. The high performance of DEXRAY suggests that image-based Android malware detectors are indeed promising. We expect our work to serve as a foundation baseline for further developing the research roadmap of image-based malware detectors. We release the dataset of apps and the extracted images to the community. We also make DEXRAY source code publicly available to enable the reproducibility of our results and enable other researchers to build on our work to further develop this research direction.

Next Steps: We expect sophisticated image representations and model architectures to offer higher performance gains compared to the studied baseline. This requires systematic ablation studies that extensively explore the techniques involved in the various building blocks of the pipeline (i.e., app preprocessing, image representation, neural network architecture, etc.) as well as the datasets at hand (e.g., with various levels of obfuscation). As part of the larger research agenda, the community should dive into the problem of interpretability of malware classification (with image representations) in order to facilitate malicious code localisation.

Data Availibility Statement. All artefacts are available online at: https://github.com/Trustworthy-Software/DexRay.

References

1. Kang, H., Jang, J.-W., Mohaisen, A., Kim, H.K.: Detecting and classifying android malware using static analysis along with creator information. Int. J. Distrib. Sens. Netw. **11**(6), 479174 (2015)
2. Petsas, T., Voyatzis, G., Athanasopoulos, E., Polychronakis, M., Ioannidis, S.: Rage against the virtual machine: hindering dynamic analysis of android malware. In: Proceedings of the Seventh European Workshop on System Security, ser. EuroSec 2014. Association for Computing Machinery, New York (2014). https://doi.org/10.1145/2592791.2592796
3. Zheng, M., Sun, M., Lui, J.C.S.: Droid analytics: a signature based analytic system to collect, extract, analyze and associate android malware. In: 2013 12th IEEE International Conference on Trust, Security and Privacy in Computing and Communications, pp. 163–171 (2013)
4. Faruki, P., Ganmoor, V., Laxmi, V., Gaur, M.S., Bharmal, A.: Androsimilar: robust statistical feature signature for android malware detection. In: Proceedings of the 6th International Conference on Security of Information and Networks, ser. SIN 2013, pp. 152–159. Association for Computing Machinery, New York (2013). https://doi.org/10.1145/2523514.2523539
5. McAfee: Mcafee labs threats report (2020). https://www.mcafee.com/enterprise/en-us/assets/reports/rp-quarterly-threats-nov-2020.pdf. Accessed 22 Feb 2021
6. Google: Android security & privacy 2018 year in review (2018). https://source.android.com/security/reports/Google_Android_Security_2018_Report_Final.pdf. Accessed 22 Feb 2021
7. Malwarebytes Lab: 2020 state of malware report (2020). https://resources.malwarebytes.com/files/2020/02/2020_State-of-Malware-Report-1.pdf. Accessed 22 Feb 2021
8. Kaspersky Lab: Kaspersky security network (2017). https://media.kaspersky.com/pdf/KESB_Whitepaper_KSN_ENG_final.pdf. Accessed 22 Feb 2021
9. Arp, D., Spreitzenbarth, M., Hübner, M., Gascon, H., Rieck, K.: Drebin: efficient and explainable detection of Android malware in your pocket. In: Proceedings of the ISOC Network and Distributed System Security Symposium (NDSS), San Diego, CA (2014)
10. Garcia, J., Hammad, M., Malek, S.: [journal first] Lightweight, obfuscation-resilient detection and family identification of android malware. In: 2018 IEEE/ACM 40th International Conference on Software Engineering (ICSE), p. 497 (2018)

11. Onwuzurike, L., Mariconti, E., Andriotis, P., Cristofaro, E.D., Ross, G., Stringhini, G.: MaMaDroid: detecting android malware by building Markov chains of behavioral models (extended version). ACM Trans. Priv. Secur. **22**(2) (2019). https:// doi.org/10.1145/3313391

12. Fereidooni, H., Conti, M., Yao, D., Sperduti, A.: Anastasia: Android malware detection using static analysis of applications. In: 2016 8th IFIP International Conference on New Technologies, Mobility and Security (NTMS), pp. 1–5 (2016)

13. Cai, H., Meng, N., Ryder, B., Yao, D.: Droidcat: effective android malware detection and categorization via app-level profiling. IEEE Trans. Inf. Forensics Secur. **14**(6), 1455–1470 (2019)

14. Wu, W.-C., Hung, S.-H.: DroidDolphin: a dynamic android malware detection framework using big data and machine learning. In: Proceedings of the 2014 Conference on Research in Adaptive and Convergent Systems, ser. RACS 2014, pp. 247–252. Association for Computing Machinery, New York (2014). https://doi.org/10.1145/2663761.2664223

15. Martinelli, F., Mercaldo, F., Saracino, A.: Bridemaid: an hybrid tool for accurate detection of android malware. In: Proceedings of the 2017 ACM on Asia Conference on Computer and Communications Security, ser. ASIA CCS 2017, pp. 899–901. Association for Computing Machinery, New York (2017). https://doi.org/10.1145/3052973.3055156

16. Szegedy, C., Vanhoucke, V., Ioffe, S., Shlens, J., Wojna, Z.: Rethinking the inception architecture for computer vision. In: 2016 IEEE Conference on Computer Vision and Pattern Recognition (CVPR), pp. 2818–2826 (2016)

17. Allix, K., Bissyandé, T.F., Klein, J., Le Traon, Y.: AndroZoo: collecting millions of Android apps for the research community. In: Proceedings of the 13th International Conference on Mining Software Repositories, ser. MSR 2016, pp. 468–471. ACM, New York (2016). http://doi.acm.org/10.1145/2901739.2903508

18. LeCun, Y., Bengio, Y., Hinton, G.: Deep learning. Nature **521**(7553), 436–444 (2015). https://doi.org/10.1038/nature14539

19. Yamashita, R., Nishio, M., Do, R.K.G., Togashi, K.: Convolutional neural networks: an overview and application in radiology. Insights Imaging **9**(4), 611–629 (2018). https://doi.org/10.1007/s13244-018-0639-9

20. Zhiqiang, W., Jun, L.: A review of object detection based on convolutional neural network. In: 2017 36th Chinese Control Conference (CCC), pp. 11 104–11 109 (2017)

21. Aloysius, N., Geetha, M.: A review on deep convolutional neural networks. In: 2017 International Conference on Communication and Signal Processing (ICCSP), pp. 0588–0592 (2017)

22. Ke, Q., Liu, J., Bennamoun, M., An, S., Sohel, F., Boussaid, F.: Computer vision for human-machine interaction. In: Computer Vision for Assistive Healthcare, pp. 127–145. Elsevier (2018)

23. Yu, D., Wang, H., Chen, P., Wei, Z.: Mixed pooling for convolutional neural networks. In: Miao, D., Pedrycz, W., Ślęzak, D., Peters, G., Hu, Q., Wang, R. (eds.) RSKT 2014. LNCS (LNAI), vol. 8818, pp. 364–375. Springer, Cham (2014). https://doi.org/10.1007/978-3-319-11740-9_34

24. Aonzo, S., Georgiu, G.C., Verderame, L., Merlo, A.: Obfuscapk: an open-source black-box obfuscation tool for Android apps. SoftwareX **11**, 100403 (2020). http://www.sciencedirect.com/science/article/pii/S2352711019302791

25. Raschka, S.: Model evaluation, model selection, and algorithm selection in machine learning, arXiv preprint arXiv:1811.12808 (2018)

26. Daoudi, N., Allix, K., Bissyandé, T.F., Klein, J.: Lessons learnt on reproducibility in machine learning based Android malware detection. Empir. Softw. Eng. **26**(4), 1–53 (2021). https://doi.org/10.1007/s10664-021-09955-7

27. Huang, T.H., Kao, H.: R2-D2: color-inspired convolutional neural network (CNN)-based Android malware detections. In: 2018 IEEE International Conference on Big Data (Big Data), pp. 2633–2642 (2018)

28. Ding, Y., Zhang, X., Hu, J., Xu, W.: Android malware detection method based on bytecode image. J. Ambient Intell. Human. Comput., 1–10 (2020). https://link.springer.com/article/10.1007%2Fs12652-020-02196-4

29. Pendlebury, F., Pierazzi, F., Jordaney, R., Kinder, J., Cavallaro, L.: TESSER-ACT: eliminating experimental bias in malware classification across space and time. In: 28th USENIX Security Symposium (USENIX Security 19), pp. 729–746. USENIX Association, Santa Clara, August 2019. https://www.usenix.org/conference/usenixsecurity19/presentation/pendlebury

30. Xu, K., Li, Y., Deng, R., Chen, K., Xu, J.: DroidEvolver: self-evolving android malware detection system. In: 2019 IEEE European Symposium on Security and Privacy (EuroS P), pp. 47–62 (2019)

31. Zhang, X., et al.: Enhancing state-of-the-art classifiers with API semantics to detect evolved android malware. In: Proceedings of the 2020 ACM SIGSAC Conference on Computer and Communications Security, ser. CCS 2020, pp. 757–770. Association for Computing Machinery, New York (2020). https://doi.org/10.1145/3372297.3417291

32. Ribeiro, M.T., Singh, S., Guestrin, C.: "Why should i trust you?": explaining the predictions of any classifier. In: Proceedings of the 22nd ACM SIGKDD International Conference on Knowledge Discovery and Data Mining, ser. KDD 2016, pp. 1135–1144. Association for Computing Machinery, New York (2016). https://doi.org/10.1145/2939672.2939778

33. Guo, W., Mu, D., Xu, J., Su, P., Wang, G., Xing, X.: LEMNA: explaining deep learning based security applications. In: Proceedings of the 2018 ACM SIGSAC Conference on Computer and Communications Security, ser. CCS 2018, pp. 364–379. Association for Computing Machinery, New York (2018). https://doi.org/10.1145/3243734.3243792

34. Palumbo, P., Sayfullina, L., Komashinskiy, D., Eirola, E., Karhunen, J.: A pragmatic Android malware detection procedure. Comput. Secur. **70**, 689–701 (2017)

35. Liu, K., Xu, S., Xu, G., Zhang, M., Sun, D., Liu, H.: A review of android malware detection approaches based on machine learning. IEEE Access **8**, 124 579–124 607 (2020)

36. Sharma, T., Rattan, D.: Malicious application detection in Android - a systematic literature review. Comput. Sci. Rev. **40**, 100373 (2021). https://www.sciencedirect.com/science/article/pii/S1574013721000137

37. Wu, D., Mao, C., Wei, T., Lee, H., Wu, K.: DroidMat: Android malware detection through manifest and API calls tracing. In: 2012 Seventh Asia Joint Conference on Information Security, pp. 62–69 (2012)

38. Burguera, I., Zurutuza, U., Nadjm-Tehrani, S.: Crowdroid: behavior-based malware detection system for Android. In: Proceedings of the 1st ACM Workshop on Security and Privacy in Smartphones and Mobile Devices, ser. SPSM 2011, pp. 15–26. Association for Computing Machinery, New York (2011). https://doi.org/10.1145/2046614.2046619

39. Kouliaridis, V., Kambourakis, G., Geneiatakis, D., Potha, N.: Two anatomists are better than one-dual-level android malware detection. Symmetry **12**(7), 1128 (2020)

40. Arshad, S., Shah, M.A., Wahid, A., Mehmood, A., Song, H., Yu, H.: SAMADroid: a novel 3-level hybrid malware detection model for Android operating system. IEEE Access **6**, 4321–4339 (2018)

41. Wang, Z., Cai, J., Cheng, S., Li, W.: DroidDeepLearner: identifying android malware using deep learning. In: 2016 IEEE 37th Sarnoff Symposium, pp. 160–165 (2016)

42. Qiu, J., Zhang, J., Luo, W., Pan, L., Nepal, S., Xiang, Y.: A survey of Android malware detection with deep neural models. ACM Comput. Surv. **53**(6) (2020). https://doi.org/10.1145/3417978

43. Karbab, E.B., Debbabi, M., Derhab, A., Mouheb, D.: MalDozer: automatic framework for Android malware detection using deep learning. Digit. Investig. **24**, S48–S59 (2018)

44. Kim, T., Kang, B., Rho, M., Sezer, S., Im, E.G.: A multimodal deep learning method for android malware detection using various features. IEEE Trans. Inf. Forensics Secur. **14**(3), 773–788 (2018)

45. Yuan, Z., Lu, Y., Xue, Y.: Droiddetector: android malware characterization and detection using deep learning. Tsinghua Sci. Technol. **21**(1), 114–123 (2016)

46. Alzaylaee, M.K., Yerima, S.Y., Sezer, S.: DL-Droid: deep learning based Android malware detection using real devices. Comput. Secur. **89**, 101663 (2020)

47. Hou, S., Saas, A., Chen, L., Ye, Y.: Deep4MalDroid: a deep learning framework for Android malware detection based on Linux kernel system call graphs. In: 2016 IEEE/WIC/ACM International Conference on Web Intelligence Workshops (WIW), pp. 104–111 (2016)

48. Wang, W., Zhao, M., Wang, J.: Effective android malware detection with a hybrid model based on deep autoencoder and convolutional neural network. J. Ambient. Intell. Human. Comput. **10**(8), 3035–3043 (2018). https://doi.org/10.1007/s12652-018-0803-6

49. Mikolov, T., Sutskever, I., Chen, K., Corrado, G.S., Dean, J.: Distributed representations of words and phrases and their compositionality. In: Burges, C.J.C., Bottou, L., Welling, M., Ghahramani, Z., Weinberger, K.Q. (eds.) Advances in Neural Information Processing Systems, vol. 26, pp. 3111–3119. Curran Associates Inc. (2013). https://proceedings.neurips.cc/paper/2013/file/9aa42b31882ec039965f3c4923ce901b-Paper.pdf

50. McLaughlin, N., et al.: Deep android malware detection. In: CODASPY 2017 - Proceedings of the 7th ACM Conference on Data and Application Security and Privacy, ser. CODASPY 2017 - Proceedings of the 7th ACM Conference on Data and Application Security and Privacy, pp. 301–308. Association for Computing Machinery Inc., March 2017. Funding Information: This work was partially supported by the grants from Global Research Laboratory Project through National Research Foundation (NRF-2014K1A1A2043029) and the Center for Cybersecurity and Digital Forensics at Arizona State University. This work was also partially supported by Engineering and Physical Sciences Research Council (EPSRC) grant EP/N508664/1.; 7th ACM Conference on Data and Application Security and Privacy, CODASPY 2017; Conference date: 22-03-2017 Through 24-03-2017

51. Nataraj, L., Karthikeyan, S., Jacob, G., Manjunath, B.S.: Malware images: visualization and automatic classification. In: Proceedings of the 8th International Symposium on Visualization for Cyber Security, pp. 1–7 (2011)

52. Oliva, A., Torralba, A.: Modeling the shape of the scene: a holistic representation of the spatial envelope. Int. J. Comput. Vision **42**(3), 145–175 (2001). https://doi.org/10.1023/A:1011139631724

53. Darus, F.M., Salleh, N.A.A., Mohd Ariffin, A.F.: Android malware detection using machine learning on image patterns. In: 2018 Cyber Resilience Conference (CRC), pp. 1–2 (2018)
54. Yadav, B., Tokekar, S.: Deep learning in malware identification and classification. In: Stamp, M., Alazab, M., Shalaginov, A. (eds.) Malware Analysis Using Artificial Intelligence and Deep Learning, pp. 163–205. Springer, Cham (2021). https://doi.org/10.1007/978-3-030-62582-5_6
55. Ünver, H.M., Bakour, K.: Android malware detection based on image-based features and machine learning techniques. SN Appl. Sci. **2**(7) (2020). https://doi.org/10.1007/s42452-020-3132-2
56. Mercaldo, F., Santone, A.: Deep learning for image-based mobile malware detection. J. Comput. Virol. Hacking Tech. **16**(2), 157–171 (2020). https://doi.org/10.1007/s11416-019-00346-7

Attacks on Visualization-Based Malware Detection: Balancing Effectiveness and Executability

Hadjer Benkraouda(iD), Jingyu Qian(✉)(iD), Hung Quoc Tran(iD),
and Berkay Kaplan(iD)

University of Illinois at Urbana-Champaign, Urbana, IL 61801-2302, USA
jingyuq2@illinois.edu

Abstract. With the rapid development of machine learning for image classification, researchers have found new applications of visualization techniques in malware detection. By converting binary code into images, researchers have shown satisfactory results in applying machine learning to extract features that are difficult to discover manually. Such visualization-based malware detection methods can capture malware patterns from many different malware families and improve malware detection speed. On the other hand, recent research has also shown adversarial attacks against such visualization-based malware detection. Attackers can generate adversarial examples by perturbing the malware binary in non-reachable regions, such as padding at the end of the binary. Alternatively, attackers can perturb the malware image embedding and then verify the executability of the malware post-transformation. One major limitation of the first attack scenario is that a simple pre-processing step can remove the perturbations before classification. For the second attack scenario, it is hard to maintain the original malware's executability and functionality. In this work, we provide literature review on existing malware visualization techniques and attacks against them. We summarize the limitation of the previous work, and design a new adversarial example attack against visualization-based malware detection that can evade pre-processing filtering and maintain the original malware functionality. We test our attack on a public malware dataset and achieve a 98% success rate.

Keywords: Malware visualization · Adversarial machine learning · Binary rewriting

1 Introduction

With the proliferation of connectivity and smart devices in all aspects of human life, these devices have become increasingly targeted by malicious actors. One

---

Benkraouda, Qian, and Tran share co-first authorship.

© Springer Nature Switzerland AG 2021
G. Wang et al. (Eds.): MLHat 2021, CCIS 1482, pp. 107–131, 2021.
https://doi.org/10.1007/978-3-030-87839-9_5

of the most common forms of attack is through malicious software (i.e., malware). According to AVTEST, one of the leading independent research institute for IT security, to date, there have been more than 100 million new malware applications in 2020 alone [19]. Inspired by the success of machine learning in other fields, researchers have proposed using machine learning for many security applications. With the rapid growth and evolving nature of new malware applications, machine learning-based solutions are a natural fit for malware detection and classification due to their robustness.

Several papers have designed malware detection systems using machine learning (e.g., [23,34]). The proposed solutions tackle malware detection from different perspectives. The main differences are in the representation used and the subsequent machine learning model selected for effective classification. These representations include raw bytes, embeddings, representative features, and binary visualization. Visualization methods, in particular, have shown high accuracy in detecting malware compared to conventional methods. These data reduction and visualization techniques for detection have shown improvements in both speed and memory efficiency [17,28]. Additionally, visualization-based techniques have achieved higher detection accuracy, mainly attributed to the applicability of deep learning techniques in detecting malware patterns [1]. We, therefore, focus our work on visualization-based malware detection models.

Nevertheless, machine learning models are susceptible to adversarial example attacks, which add imperceptible non-random perturbations to test samples, causing machine learning models to misclassify them. Successful adversarial examples have been seen to fool systems into misclassifying people [36], cause systems to recognize street stop signs as speed limit signs [10], or cause voice-controllable systems to misinterpret commands or perform arbitrary commands [40].

Recent work has shown that machine learning-based techniques for malware detection are also susceptible to adversarial examples. In these systems, the attacks alter the malware binaries to cause the target model to classify the malware sample as benign or vise versa. However, adversarial examples in this domain are more challenging to produce. In addition to the constraint of imperceptibility and minimal changes that conventional adversarial examples must comply with, adversarial examples for malware detection must maintain the original malware functionality. This means that the attacker cannot change the bytes arbitrarily. Instead, the attacker has to understand the functionality of the malware and perform careful changes.

There have been previous attempts to create adversarial examples against visualization-based malware classifiers [21,27]. These attacks produce adversarial examples either by using conventional image-based techniques such as the Fast Gradient Sign Method [15] or Carlini and Wagner method [4] or by injecting byte values to unreachable regions within the binary. These attacks are simplistic and can be detected and removed easily with minimal countermeasures [24].

In this paper, we propose a new adversarial example attack that combines binary rewriting and adversarial attacks in image classification. We target a convolutional neural network (i.e., CNN) model for malware detection. Because there is no open-sourced code for visualization-based malware detection, our first phase of the project includes constructing the malware detection model (Fig. 1 left). We apply a similar CNN structure as previous work for visualization-based malware detection and achieves an overall accuracy of 99%. In the second phase of the project (Fig. 1 right), we design an adversarial example attack against this malware detection model. Our attack performs additive changes to the original malware and ensures that the added instructions are semantic NOPs, i.e., they do not change values of any register or manipulate the program state. Our attack achieves a 98% success rate on a public malware dataset. The success of the proposed attack reveals that it is necessary for visualization-based malware detection to perform more advanced and robust protection against adversarial examples other than simply filtering the padding or the non-reachable header section.

Phase 1: Construct the malware detection model

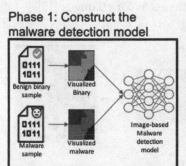

Phase 2: Design algorithms to generate robust adversarial examples

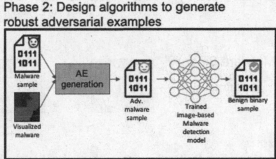

Fig. 1. The two project phases: constructing the malware detection model and designing the adversarial example attack.

The rest of the paper is organized as follows. In Sect. 2, we introduce background and related work on visualization-based malware detection and adversarial machine learning. In Sect. 3, we provide our detailed design of the attack against visualization-based malware detection and illustrate how we solve the challenges of creating successful adversarial examples while maintaining the original malware functionality. In Sect. 4, we discuss our experiment setup and measure our attack success rate. In Sect. 5, we discuss limitations of our attack and potential future work.

2 Background and Related Work

In this section, we introduce the background and related work on visualization-based malware detection. We then discuss some traditional methods to camouflage malware, adversarial machine learning and how attacks against image classification and malware detection work. Finally, we include a table for SoK of the papers that we mentioned, and point out their limitations.

2.1   Malware Visualization

With the development of image processing technology, visualization-based techniques are also proposed for malware detection and analysis. These techniques can be applied directly to the binary without complicated disassembly and execution process. Researchers have proposed approaches to visualize malware as gray-scale or RGB-colored images. From these images, machine learning techniques, such as CNN, can classify whether the tested software is benign or malicious.

Figure 2 illustrates a typical approach to visualize the malware as an image. The malware binary is grouped by 8-bit vectors. Each vector represents a value from 0 to 255, which can be mapped to a gray-scale pixel value. The shape of the final image depends on the width of the image, which is usually a tunable parameter, and the size of the malware binary in bytes. This methodology can be adapted to visualize the malware as an RGB-colored image, which considers different feature types and represents them in different color channels [12].

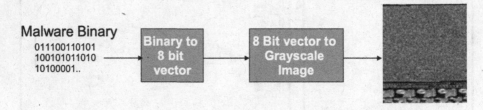

Fig. 2. Typical approach to visualize the binary in gray-scale [30]

Here we introduce a few projects from the literature that used several different malware visualization techniques. Han et al. [17] proposed a malware visualization method that converts the binary to a gray-scale bitmap image and then generates the entropy graph for the entire malware binary. They then used a histogram similarity measuring method to group malware within the same malware family. Nataraj et al. [30] also visualized the malware binary as a gray-scale image but extracted texture features to characterize and analyze the malware. They used GIST to capture texture features, and apply the K-nearest neighbors algorithm with Euclidean distance to classify the malware into different malware families. Xiaofang et al. [39] mapped malware binaries to gray-scale images, extracted a 64-dimension feature vector from the image, and performed fingerprint matching to identify similar malware.

Unlike the other work, which converts the malware binary to a gray-scale image, Fu et al. [12] took a different approach to visualize the malware (Fig. 3). Their approach extracted both local and global features to generate an RGB-colored image for the malware. Specifically, they extracted three types of features, including section entropy, raw byte values, and relative size of each section to the whole file. For raw byte values, they use the same approach to visualize malware as gray-scale images (Fig. 2). Each types of features occupies a single color channel (i.e., either red, green, or blue). For the final classification process, they compared different machine learning techniques, including random forest, K-nearest neighbors, and support vector machine.

Han et al. [18] proposed a novel method to classify malware and malware families. They extracted opcode instruction sequences from the malware binary as features and generated the RGB-colored image matrix from these features. The image matrices are compared with each other using selective area matching [18] to group malware into malware families.

Another work interestingly extends the field by visualizing the behavior of the malware instead of the malware binary itself, and suggests that any feature of a malware can be visualized to determine its classification [35]. This work further indicates that the possibility of malware visualization is limitless as a program has a lot of useful features ranging from its behavior to its metadata. But, for this work, it specifically focuses on the malware behavior by running an API call monitoring utility through a Virtual Machine (i.e., VM) to examine the APIs used while the program is executed in user mode [35]. While there are several other techniques to capture malware behavior, such as capturing the network activity of the malware, or the changes in the operating system's resources, API monitoring has been chosen in this study due to its conciseness and the shortcomings of other techniques that have been discussed in detail [35]. Afterwards, the calls are mapped to hot colors, specifically such as red, or orange, for classification. Finally, the classification is mostly done through a similarity ratio to the tested software against the known malware's color mapping [35].

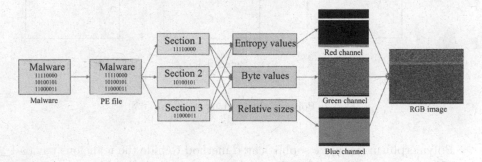

Fig. 3. Visualize the malware as an RGB image, considering both local and global features [12]

2.2 Traditional Malware Camouflage

Although there are several methods of malware detection based on visualization, attackers can still employ various methods such as encryption and obfuscation to hide their malware in the targeted software's code and counter static malware detection methods [7, 26, 33].

Malware encryption intends to encrypt the malware body to hide its intentions and avoid static analysis detection so that a direct signature matching defense cannot detect the malware [7,20,26]. It relies on a decryption loop (a.k.a., decryptor) to decrypt the malicious payload and execute the malware. Therefore, if the defense can find out the decryption loops in the malware, it can decrypt the malicious code first and then perform a simple signature matching to detect the malware. In addition, the original defense can be easily augmented with a signature checking component to identify suspicious decryption loops within the malware. Visualization-based defense can be even better at detecting malware encryption because it can extract locality features specific to suspicious decryption loops of the malware. Even a more complicated malware encryption technique that picks different decryptors for different malware (i.e., oligomorphism) only prolongs the detection time [33].

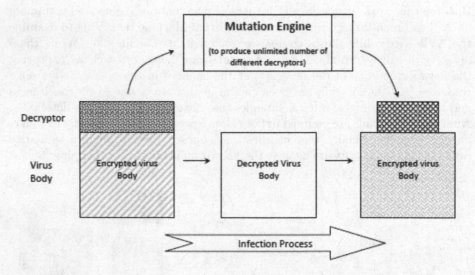

Fig. 4. Polymorphism virus structure [33].

Polymorphism is a more sophisticated method to hide the malicious payload based on malware encryption (Fig. 4). It uses several types of transformations on the decryptor, such as changing the order of instructions with additional jump instructions to maintain the original semantics and permuting the register allocation to deceive anti-virus software [7]. It also typically injects junk or dead codes to further mutate the decryptor so that it is hard to recognize it [33]. However, after enough emulation and a simple string matching algorithm, the underlying encrypted sections of the malware can still be revealed [33].

A major drawback of either malware encryption or polymorphism is that it relies on an explicit decryptor to decrypt the malicious payload and execute the malware. This leaves a significant mark on the malware that can be relatively easily detected. On the other hand, metamorphism is a more advanced

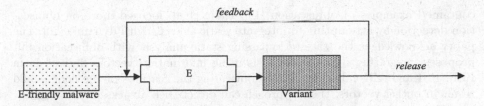

Fig. 5. Metamorphism structure: a metamorphic engineer is responsible for changing the malware instructions to equivalent ones probabilistically [33].

technique to camouflage the malware without using any encrypted parts. Malware metamorphism is a technique to mutate the malware binary using different obfuscations by a metamorphic engine (Fig. 5). In this way, the attacker changes the syntax of the original malware but keeps the original malware behavior. In particular, metamorphism allows the malware to change its opcode with each execution of the infected program. Alam et al. [2] group some typical obfuscations used in metamorphism into three categories. The first category is the opcode level obfuscation, which includes instruction reordering, dead code insertion, and register renaming [2]. The second category is control flow level obfuscation, which includes changing the order of instructions, and applying branch functions, opaque predicates, jump tables, and exception tables [2]. The last category is obfuscation by self-modifying code, which intends to change instructions during runtime in order to hide malicious payload to avoid reverse engineering and detection by anti-malware software [2].

It has been shown that malware metamorphism can easily defeat the signature-based malware detection [6] because signature-based detection is unable to capture the changes of the malware due to dynamic code obfuscation. However, metamorphic malware is usually initiated from known malware, and with the initial knowledge of existing malware, it is still possible to detect malware metamorphism. Zhang et al. [41] proposed a defense to characterize the semantics of the program and perform code pattern matching, based on static analysis of control and data flow of call traces. Alam et al. [2] proposed a metamorphic malware analysis framework that builds the behavioral signatures to detect metamorphic malware in real-time. Chouchane et al. [5] proposed another method to detect the existence of a metamorphic engineer by checking the likelihood of a piece of code generated by some known metamorphic engines. In addition, the metamorphic malware typically has a significant proportion of binary related to the metamorphic engine, which can be recognized by a more advanced detector, such as a visualization-based malware detector.

Go et al. discussed the importance of developing new approaches against polymorphism and metamorphism specifically [13]. Their method converts the binary to a grey-scale image and uses the ResNeXt CNN model to build resiliency against such malware camouflage techniques [13]. Their paper does not explicitly discuss their method's effectiveness against obfuscation attacks [13]. The researchers used the Malimg dataset but did not mention that their dataset contained examples of obfuscation [13]. Islam et al. focused more on obfuscation detection by attempting to integrate static analysis with dynamic [20]. The paper acknowledges the ease of bypassing static analysis with obfuscation but proposes integrating dynamic analysis using information vectors derived from FLF, PSI and API calls [20]. Since obfuscating the features of the code would result in outlier vectors, their approach can detect such attacks [20].

2.3   Adversarial Machine Learning

In this section, we introduce adversarial machine learning. In general, adversarial machine learning aims to fool the machine learning model by carefully generating adversarial examples through evasion attacks or polluting the training phase through poisoning attacks. We focus our discussion on evasion attacks leveraged against image classification and malware detection due to their many close similarities to visualization-based malware detection.

Attacking Image Classification. Previously, image classification has been used in many applications not related to binary classification. In these fields, multiple attacks have been produced to cause errors in detection. Goodfellow et al. [15] illustrated the fast gradient sign method (i.e., FGSM) to generate adversarial examples. Figure 6 shows how FGSM is applied to cause the image classifier to mistakenly classify a panda to a gibbon by adding carefully crafted perturbation. In Carlini et al.[4], it was shown that the addition of random noise to images can significantly reduce the accuracy of classifiers while being imperceptible to the human eye. These attacks can be mitigated to some degree by denoising techniques, as proposed in Liao et al.[25]. Nevertheless, such mitigation efforts do not fully reverse the perturbations and may introduce more noise accidentally. Furthermore, as shown in Eykholt et al.[11], classification can also be interrupted by altering small sections of an image, where the image would still be readable by a human eye.

A notable difference in image classification for binary visualization is that the validator for images lies in code execution instead of human recognition. As a result, perturbations intended for adversarial attacks on binary visualization must continue to function as an unchanged code sample. A code sample that maintains functionality after modification can be said to maintain executability, as the code will execute as intended.

Attacking Malware Detection. Because malware detection from raw bytes relies heavily on the performance of the machine learning model for classifying the image embedding of the malware (e.g., RGB-colored image), it is also vulnerable to similar attacks against image classification. However, there are more difficulties in generating valid adversarial examples in the malware detection domain than in the image classification domain. The main challenge is to solve the inverse feature-mapping problem [32]. Pierazzi et al. [32] proposed a novel formalization of problem-space attacks and a novel problem-space attack in the

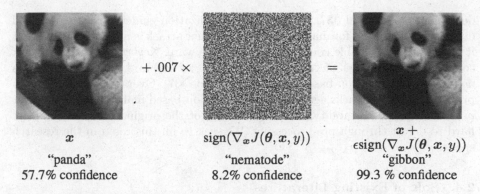

x

"panda"

57.7% confidence

$\mathrm{sign}(\nabla_x J(\theta, x, y))$

"nematode"

8.2% confidence

$x +$
$\epsilon\mathrm{sign}(\nabla_x J(\theta, x, y))$

"gibbon"

99.3 % confidence

Fig. 6. Adversarial example generated using FGSM [15].

Android malware domain. They used conditional statements that are never executed during runtime to wrap the malicious code payload. In particular, they used opaque predicates to ensure the obfuscated conditions always resolve to false but look legitimate. They showed that their attack against Android is successful against the Drebin classifier [3] and several SVM-based malware detectors.

Liu et al. proposed introducing perturbations to the visualized binary to lower the success rate of ML-based malware detectors [27]. The paper introduced a method that leverages gradient descent and L-norm optimization methods. However, as it changes the image in potentially unexpected ways, it cannot guarantee the executability of the perturbed malware.

Khormali et al. [21] showed simple adversarial examples to bypass visualization-based malware detection. Their attack only attempts rudimentary methods to generate adversarial examples, namely padding and injection [21]. It preserves the executability and functionality of the original malware, but it is easy to detect and not scalable. Kolosnjaji et al. [22] proposed a gradient-based attack to evade ML-based malware detection. Their attack also injects padding bytes to the original malware but does not consider preserving the functionality and does not target visualization-based malware detection. Demetrio et al. [9] proposed a general framework to perform white-box and black-box adversarial attacks on learning-based malware detection by injecting the malicious payload to the DOS header. However, the authors also claimed that their header attacks could be easily patched through filtering process before classification. Grosse et al. [16] demonstrated the attack against a deep neural network approach for Android malware detection. They crafted adversarial examples by iteratively adding small gradient-guided perturbation to the malware on application level instead of directly perturbing the binary. They restrict the perturbation to a discrete set of operations that do not interfere with the malware functionality. In addition, they discussed some remedies against their attacks, including feature reduction, distillation, and adversarial training (i.e., re-training the model with the addition of adversarial examples). Their work focused more on attacking application-level malware detection instead of visualization-based malware detection. Sharif et al. [37] proposed an optimization-guided attack to mislead deep neural networks for malware detection. Their attack is more invasive in that it changes the reachable code section of the malware. Nevertheless, their attack considers a limited set of transformation types of malware functions and does not target visualization-based malware detection [37]. Overall, there is no work proposing robust attacks against the visualization-based malware detection that preserve both executability and functionality of the original malware and are hard to detect through pre-processing. We seek to fill this room in the research space.

2.4   SoK of Existing Literatures

Finally, we provide Table 1 as our SoK of existing malware visualization techniques and attacks against them. For each reviewed work, we summarize its methodology and list its limitations.

3   Robust Adversarial Example Attack Against Visualization-Based Malware Detection

In this section, we focus on the workflow of generating adversarial examples, and we leave the construction of the malware detector to Sect. 4.1. We have two main goals for our adversarial example attack. Firstly, we aim to find an adversarial example generated from a single malware such that the malware detector will misclassify it as benign software. Secondly, an adversarial example generated in this way must maintain the functionality of the original malware. An overview of the full workflow of our adversarial example generation algorithm is shown in Fig. 7. At a high level, the attack starts with using a mask generator to add empty spaces to instruction boundaries where perturbations are allowed. Then, the adversarial example generator (i.e., *AE generator*) will generate the optimal adversarial example in the image space. To ensure that the original malware functionality is not changed, we use a NOP generator to produce a semantic NOP list and update the optimal adversarial example to the closest matching viable one that preserves malware functionality. If this processed adversarial example is still misclassified as benign, then our attack succeeded. Otherwise, we relaunch the AE generator, starting from the failed adversarial example, creating a new optimal AE, and starting a new iteration. We iterate until we produce a successful adversarial example or we reach a pre-set threshold of iterations. In the following sub-sections, we discuss the details of each component in the workflow.

3.1   Mask Generator

The first step of our attack workflow aims at controlling and locating where the perturbations can be added. This step is provided to ensure both executability

Table 1. SoK of reviewed papers: malware visualization and adversarial attacks against malware visualization software.

Category	Paper	Methodology	Limitations
Malware visualization on binary: Gray-Scale	Han et al. [17]	Converts binary to the bitmap image and generates the entropy graph from visualized malware	Hard to classify packed malware binaries
	Nataraj et al. [30]	Extracts image texture features from visualized malware	Relying on global texture features can be beat by attackers
	Xiaofang et al. [39]	Extracts a 64-dimensional feature vector and performs fingerprint matching to identify similar malware	Relying on global image features can be beat by attackers
Malware visualization on binary: RGB-Colored	Fu et al. [12]	Combines entropy, relative section size, and raw bytes to generate an RGB-colored image	Limited to PE format
	Han et al. [18]	Extracts opcode instruction sequences	Classification is not yet automated
Malware visualization on behavioral features	Shaid et al. [35]	API call monitoring	Does not consider network behavior and does not work directly on malware binary
Attacking malware detection	Pierazzi et al. [32]	General problem-space attack for inverse feature-mapping	Attacks focus on Android malware and are not against neural network-based detector
	Liu et al. [27]	Gradient descent to perturb binary	Not guarantees executability
	Kormali et al. [21]	Padding and injection	Easy to detect and not scalable
	Kolosnjaji et al. [22]	Padding and injection	Does not preserve functionality and does not target visualization-based malware detection
	Demetrio et al. [9]	Injects the malicious payload to DOS header	Easy to patch through filtering process before classification
	Grosse et al. [16]	Iteratively adds small gradient-guided perturbations	Only targets Android malware and does not attack on binary level
	Sharif et al. [37]	Manipulates code section guided by an optimization function	Considers only a limited set of manipulation types

and robustness to simple pre-processing defenses while maintaining the original semantic operation. The central intuition of this step is to allow additional instructions to be embedded within the code section so that they are not easily distinguishable from the rest of the code section and, in the meantime, ensure that these added instructions do not introduce any changes to the original mal-

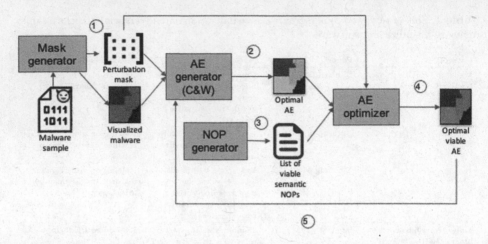

Fig. 7. The overview of the workflow of the adversarial example generation algorithm.

ware instructions. These additional instructions will serve as empty spaces to allow perturbing the malware in the image representation.

To achieve this, we create a *mask generator* (Fig. 8). The algorithm of the *mask generator* is as follows. First, we extract the code section of the malware sample and identify the instruction boundaries. Next, we need to decide the size of each perturbation block, its location, and frequency. The attacker can set these parameters to satisfy certain size limitations of the perturbation relative to the original malware size to make it harder to detect. On the other hand, the attacker can also increase the frequency of the perturbation blocks to make the attack easier. The perturbations can be initialized to random inputs or naive NOPs. After that, these perturbations are inserted into the expected instruction boundaries. In this way, we make sure that the original malware instructions are not changed at all because we never add extra bytes to the middle of the binary of any instruction. With this malware augmented with the perturbation sequences initialized to naive NOPs, we use a binary-to-image converter to represent the malware in the image space. The binary-to-image converter treats the consecutive 8 bits of the binary as a single pixel to build a PNG file. Figure 8 illustrates how our mask generator adds empty space to allow perturbation. In this example, we add a single NOP every two instructions. The image representation of the binary is expanded by two pixels (i.e., the pixel marked with '*') due to the two added nops.

Besides the malware image, the *mask generator* produces a mask in the form of an array of the same dimension as the augmented malware image. The mask flags the locations where perturbations are allowed with ones, while the rest of the array is filled with zeros. We name this mask the *perturbation mask*.

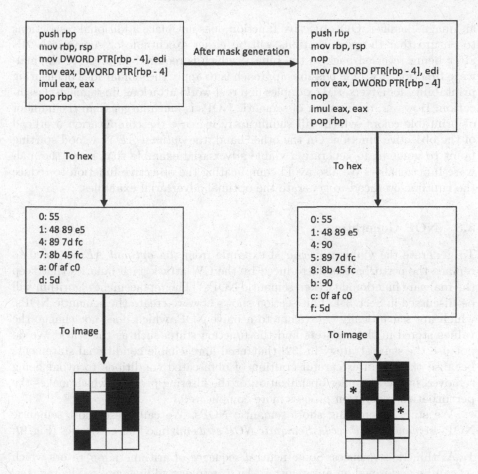

Fig. 8. Mask generator adds empty space to allow perturbation

3.2 AE Generator

Once the *perturbation mask* and the augmented malware image are generated, we launch a modified version of the CW attack [4] to generate the optimal adversarial example (i.e., *optimal AE*) in the image space, which is misclassified as benign by the malware detector. The only difference is the application of the *perturbation mask* to further restrict the positions of perturbations. The objective function is given as

$$\min \|M\delta\|_2 + C \cdot f(x + M\delta) \quad s.t. \quad x + M\delta \in [-1, 1]. \tag{1}$$

Here M is the *perturbation mask*, and x is the augmented malware image produced by the *mask generator*. The *optimal AE* is unlikely to correspond to the perturbed malware that maintains the original malware functionality because the CW attack only intends to attack the malware detector, which is modeled as an image classifier. Our objective function does not place additional restrictions to ensure that the perturbations will be either executable or semantic NOPs after being reversed back to the binary, which is required to preserve the malware functionality. One possible approach is to apply the similar idea of ensuring printability for adversarial examples in a real-world attack against image classification. However, the number of semantic NOPs is much larger than the number of printable colors, which will significantly increase the computation overhead of the objective function. On the other hand, the *optimal AE* is a good starting point to guide us to generate a viable adversarial example that keeps the malware functionality. We also avoid complicating the objective function to reduce the runtime overhead to generate the optimal adversarial examples.

3.3   NOP Generator

To generate the viable adversarial example from the *optimal AE*, we need to replace the perturbations introduced by the CW attack with binaries that keep the malware functionality (i.e., semantic NOPs). The replacement algorithm will be discussed in Sect. 3.4. This section shows how we create the semantic NOPs, which are semantically equivalent to a naive NOP, which does not change the values stored in the registers and the function state, such as the stack. We do not use the same strategy in [32] that used unreachable conditional statements because that requires careful crafting of obfuscated conditions to avoid being removed by the compiler optimization or the filtering process, which makes the perturbation generation process more complicated.

We start by creating short semantic NOPs. We call these short semantic NOPs *semantic NOP seeds*. *Semantic NOP seeds* fall into four categories (Fig. 9):

1. Arithmetic sequences: Some neutral sequence of arithmetic operators which can be performed on any register. An example is adding zero to the register.
2. Movement sequences: Instructions to move the register value back and forth or jump to the defined locations to skip the regions that are not expected to execute. Examples are moving the register value to itself or jumping to the immediate next instruction.
3. Logical sequences: Some neutral sequence of logical operators which can be performed on any register. An example is ANDing 1 to the register.
4. Other miscellaneous sequences: A sequence of simple NOPs or a sequence to change and recover the flags.

Because the perturbation space is not pre-determined, we do not generate semantic NOPs for any arbitrary size. Instead, we build the *NOP generator* to combine *semantic NOP seeds* to make longer semantic NOPs that can fit larger perturbation spaces. For instance, if the *NOP generator* is asked to compute 3-byte semantic NOPs and the naive NOP (i.e., byte 90 in heximal) is one of the *semantic NOP seeds*, then it can combine three NOPs. Given the expected size of each perturbation block, the *NOP generator* produces a list of semantic NOPs of the given size.

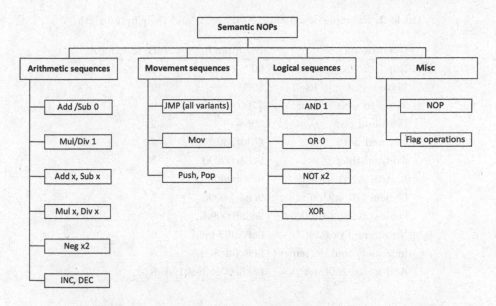

Fig. 9. Semantic NOP seeds used to construct longer semantic NOPs.

It is necessary to keep the byte length of *semantic NOP seeds* as small as possible so as to improve the ability for the *AE optimizer* to produce a viable adversarial example that maintains the malware functionality. However, too much restriction on the byte length also limits our ability to generate enough *semantic NOP seeds*. In our design, we pick the minimum size of a *semantic NOP seed* as one byte (i.e., the byte length for a naive nop) and the maximum size of a *semantic NOP seed* as eight bytes. We provide our example *semantic NOP seeds* and the corresponding byte size in Table 2. We admit that this is a simple design choice and is far from the optimal selection. We also do not comprehensively list all the possible operations given the operation type, which can make the set of *semantic NOP seed* too large. However, our evaluation results reveal that this selection is already enough for us to generate adversarial examples for malware that maintains functionality with high success rate.

3.4   AE Optimizer

In this step, we build a module, the *AE optimizer*, which produces a viable adversarial example that maintains the original malware functionality. The *AE optimizer* takes in the *perturbation mask*, the *optimal AE* generated from the CW attack, and the list of semantic NOPs produced by the *NOP generator*. Next, the *AE optimizer* locates the allowed positions for perturbations in the *optimal AE* using the *perturbation mask*. Subsequently, the Euclidean distance between the instruction sequences in the allowed perturbation spaces and the semantic NOPs is calculated using the following equation:

Table 2. Example of semantic NOP seeds and their byte length.

Operation Type	Example in Hex	Byte Length
Nop	90	1
Move register to itself	89c0	2
Jump to next instruction	7700	2
Push and pop	5058	2
Not and not	f7d0f7d0	4
Add/subtract 0	9c83c0009d	5
Logical AND with 1	9c83e0ff9d	5
Logical OR with 0	9c83c8009d	5
Logical XOR with 0	9c83f0009d	5
Negate and negate	9cf7d8f7d89d	6
Increment and decrement	9cffc0ffc89d	6
Add x and subtract x	9c83c00183e8019d	8

$$d\left(p,q\right) = \sqrt{\sum_{i=1}^{n}\left(q_i - p_i\right)^2} \tag{2}$$

Here p and q are the generated instruction sequence and the semantic NOPs. This process identifies the semantic NOPs closest to each of the sequences in the allowed perturbation space of the *optimal AE*. In the current implementation, this process is done sequentially for each perturbation block; however, it can be easily parallelized to improve the runtime performance. After that, the semantic NOPs with the minimum distance are used to replace the perturbation blocks in the *optimal AE*. The new adversarial example is called the *optimal viable AE*.

Finally, we pass the *optimal viable AE* to our malware detector for classification. If it is classified as benign, we stop the process because we have already produced a successful adversarial example. If it is correctly classified as malware, it will be used as the starting point for another iteration of the CW attack, and the process is repeated. We expect that starting from a failed optimal viable AE can direct us better to the successful optimal viable AE. However, it is possible that the AE can be stuck in the local optimum. Another more general approach is to start over the whole process again from the visualized malware augmented with randomly initialized semantic NOPs.

4   Evaluation

4.1   Experiment Setup

Malware Detection Model. We use a CNN-based malware detector as our attack target. Because there is no open-source code for the visualization-based malware detector, we build our own CNN by following a similar structure from previous work. Specifically, we re-built the structure described in [21, 22]. Our CNN is composed of an adaptive average pooling layer to handle inputs of different dimensions, two consecutive convolutional layers with max-pooling, and three fully connected layers. We use ReLU as the activation function (Fig. 10).

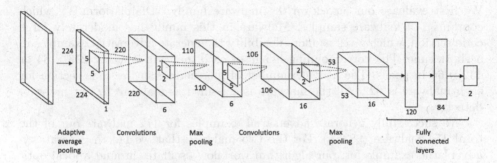

Fig. 10. Malware detection neural network model structure.

Dataset. We build our dataset by combining one public malware dataset and one public benign software dataset. The malware dataset is from UCSB's public malimg dataset [29], which has a collection of around 9500 malware formatted as PNGs. The benign dataset is a collection of around 150 benign software from the Architecture Object Code Dataset [8]. The software was chosen from the AOCD with various functionalities to ensure that the classifier did not learn latent features representative of the type of benign code.

Model Training. A relatively small subset of the malimg dataset is used to acquire a 50% split between malware and benign images in the training data. This subset was retrieved randomly from the malimg dataset, and examples were chosen without considering their corresponding malware family. In doing this, we prevent the classifier from learning any latent features from any particular malware family and improve its accuracy for any malware class in the validation set. Validating this model's accuracy, we are able to confirm the classification accuracy of state-of-the-art models, with a 100% accuracy on identifying benign software and a 99% accuracy on correctly identifying malware samples.

Attacking the Malware Detector. We evaluate our attack using the malware from the malimg dataset [29]. For each malware, we augment it with empty spaces per instruction initialized with eight naive NOPs. If the *AE generator* fails to produce the *optimal AE*, we consider it an attack failure. If the *AE generator* can produce the *optimal AE*, we run the *AE optimizer* to replace perturbations with the closest semantic NOPs. We set the iteration threshold to be ten. If the *optimal viable AE* is generated within the ten iterations of the CW attack and AE optimization and can be successfully misclassified as benign, we view it as a successful attack.

4.2   Results

We first evaluate our attack on the malware family, "Dialplatform.B", which contains 174 malware samples. Malware in this family has moderately large code section, which gives us more flexibility to generate adversarial examples in a short time. The corresponding image of each malware is either 216×64 or 432×64 in pixels. The average number of instructions in the code section for each malware is 622. All the malware is classified as malware by our malware detector.

We successfully generate adversarial examples for 172 malware out of the total 174 malware (98.9%). For the two malware that we fail to generate an adversarial example for, our algorithm workflow oscillates around a local optimum, so our iteration of CW attack and AE optimization (steps 2, 4, 5 in Fig. 1). There are two potential methods to avoid the local minimum oscillation issue. First, we can initialize the empty spaces between instructions with random semantic NOPs. Randomized initialization has already been applied frequently to solve similar local optimum oscillation problems before. Second, in the AE optimizer step (Sect. 3.4), we could pick a sub-optimal viable adversarial example if we detect local optimum oscillation. In this way, our algorithm can break the local optimum while also searching for a valid adversarial example.

For all of the adversarial examples, the functionality of the original malware is preserved by construction since we only instrument the original binary with semantic NOPs at instruction boundaries. All of the adversarial examples can be generated within five iterations of the CW attack and AE optimization. The running time to generate the adversarial example is given in Fig. 11. On average, it takes 77.8 s to generate the adversarial example for the malware with around 600 instructions, and the time overhead is in AE optimization step. The expansion rate of the original malware due to augmentation is shown in Fig. 12. On average, the perturbation size is 35.82% of the size of the original malware as seen from Fig. 12. We argue that though the expansion rate is high, the added perturbation is still hard to filter out due to the flexibility of our semantic NOPs. The attackers can even build a much larger semantic NOP set than our current version. They can also consider using dead code branching. In this way, unless the users know the semantic NOPs and the initial size of the malware, it is hard to filter out our attacks.

We achieve similar results for other malware families with similar sizes of the code section. On the other hand, our algorithm does not work well for malware with very small code sections. We argue that the spaces to add semantic NOPs are too few and too small to allow enough perturbation to cause misclassification due to the small code section. We expect that enlarging the mask size can improve the attack success rate, but this might defeat the attacker's purpose for distributing small and easy-to-transmit malware. Another potential way is to combine perturbations in code sections with those in other sections, such as data sections. However, we argue that keeping executability can be hard and naive padding is relatively easy to filter out. We leave attacks for malware with a small code section as future work.

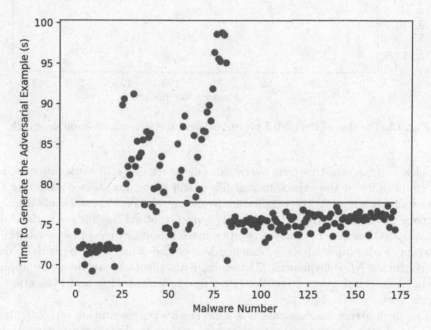

Fig. 11. The runtime evaluation for the adversarial example attack.

To further test the end-to-end methodology for generating adversarial examples for malware with a larger number of instructions, we also run the same experiment with 11 malware from the "Lolyda.AA3" class. Each malware contains at least 200,000 instructions. We achieve an attack success rate of 81.8% (i.e., 9 out of 11). The two failures are also due to local optimum oscillation issues that we face when generating adversarial examples for malware in "Dialplatform.B" family. We expect random initialization and sub-optimal AE optimizer can solve the problem and leave it to future work. On the other hand, the major problem for generating adversarial examples for large malware is the time overhead. In our experiments, the time overhead reaches about six hours to generate a single adversarial example. As a proof-of-concept attack and assuming the

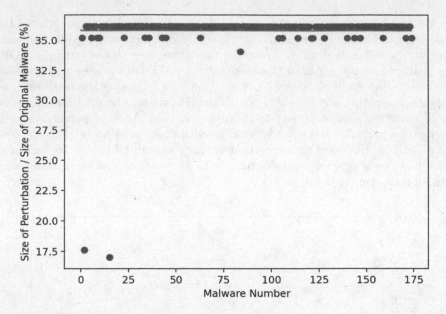

Fig. 12. The size of the added perturbation of the adversarial example attack.

attacker can tolerate this time overhead, our attack can still work properly, but the practicality of the attack in real life is still under question considering the size overhead added to the original malware as well. We expect parallelism can improve the running time. Specifically, our current AE optimizer finds viable semantic NOP sequences for each empty space defined by the mask sequentially. However, each empty space is independent of the other if we perturb it with self-contained NOP sequences. Therefore, parallelization can be easily applied in the AE optimization step. We leave its implementation and evaluation as future work.

Our high attack success rate is a conservative representation of how vulnerable visualization-based malware detection can be to the adversarial examples. In our experiment, we set each perturbation block to be small (i.e., 8 bytes). A more powerful attacker who would like to risk being detected more can further increase the size of each perturbation block so that the attack can be easier to launch.

5 Discussion

5.1 Limitations

While the proposed attack can successfully find adversarial examples with a high success rate, due to the nature of the optimization algorithm, the time necessary to find the *optimal viable AE* increases drastically with the size of the original malware. In our evaluation, we performed adversarial example attacks on the

"DialPlatform.B" malware family, where each malware image is of dimension 216×64 or 432×64 with an average of 622 instructions. Since the images are of reasonably small dimensions, the potential perturbation spaces are fairly limited. As a result, an adversarial example can be generated in less than two minutes. However, for larger malware, as in the "Lolyda.AA3" family, producing a single optimal adversarial example can take a few hours. As each perturbation block can be independently replaced with the closest semantic NOPs, we expect parallelization to improve the running time.

In addition, our attack only adds instruction to the malware's code sections. Therefore, when the code section for the original malware is small, it will be hard to provide enough perturbations to cause the classifier to misclassify the malware image. Similar issues occur when the data section is much larger than the code section. One possible approach to find an easier way to generate perturbation is to find out the hot zone for the machine learning detector on the malware in the first place and then only adding enough semantic NOPs to these locations. Another approach to solving this challenge is to design a mechanism to perturb the data section without affecting the original functionality of the malware.

There are some potential defenses against general attacks to malware detection. Tong et al. [38] proposed a method to boost the robustness of feature-space models by identifying the conserved features and constraining them in adversarial training. Their defense is evaluated only on PDF malware, and we plan to evaluate further the defense on more malware families and more diverse model types.

5.2 Future Work

The proposed attack algorithm is a proof-of-concept and a work-in-progress, and there are several research directions we would like to explore as future work:

1. In the current version of the attack, the size and frequency of the added perturbations are chosen beforehand by the attacker. In the development of the attack, we would like to explore the possibility of adding these two parameters (i.e. size and frequency of the perturbations) to the optimization process. We speculate that this can improve the performance of our AE generation algorithm. First, it can lead to faster convergence into a viable AE. Additionally, it can also lead to a smaller perturbation size that is customized for each malware sample.
2. Another avenue of improvement to our baseline algorithm is in speed performance as the baseline algorithm does not leverage any speed optimizations. Our AE generation algorithm can allow batching and parallelization to enhance the speed performance. First, we plan to batch instructions to be processed at once instead of processing individually. Additionally, since Euclidean distance calculations do not have to be done sequentially we plan to calculate all the distances in parallel.
3. Another avenue that we would like to explore is the intrinsic properties of NOPs. Specifically, we would like to study the difference between NOPs and

whether some NOPs are better than others. Additionally, we would like to draw from the field of software engineering in creating semantically similar code using modifications that do not change the semantics at a higher-level language (e.g. adding empty loops or if statements that would never be true).

4. In our current implementation, we restrict adding perturbation to the code section. In the algorithm, we would like to explore the effects of adding perturbations to other sections and understanding its effects on executability, robustness to pre-processing, and maintaining semantic operations.

5. We plan to perform a comprehensive comparison between our proposed attack with the state-of-the-art attacks with respect to speed, stealthiness, and deviation from the original binary. Additionally, we want to evaluate our attack's success against the defense and detection mechanisms available (e.g. Virus-Total).

6. To have an end-to-end solution we would like to add a module that checks executability of each of the binary (e.g., using *execve* command).

7. Our current work has already revealed that malware detection based on binary visualization can be beaten by an adversarial example with a high success rate. In the meantime, the attacker can also maintain the functionality of the original malware. Therefore, our work motivates future directions to propose defenses against our attack. Previous defenses against adversarial examples, such as defensive distillation [31] and adversarial training [14], does not usually focus on real-world tasks, such as defending malware detection models. Whether these defenses are effective in this use case is worth exploring.

6   Conclusion

In this work, we provide a literature review on existing techniques and attacks for malware detection. We summarize their limitations, and propose a novel end-to-end method for generating adversarial examples against visualization-based malware detection. We design our attack workflow to ensure that the malware sample remains executable while maintaining its original functionality. Additionally, we ensure that the added perturbation is robust against pre-processing by inserting semantic NOPs in the reachable code section. We construct a dataset that includes both malware and benign software samples and use it to build an visualization-based ML malware detector that achieves high accuracy. Next, we design a workflow that generates semantic NOP sequences and use them to construct viable adversarial examples. Our results show that it is possible to successfully generate adversarial examples that can bypass a highly accurate visualization-based ML malware detector while maintaining executability and without changing the code operation. Our work motivates the design for more robust visualization-based malware detection against carefully crafted adversarial examples.

References

1. Abuhamad, M., Abuhmed, T., Mohaisen, A., Nyang, D.: Large-scale and language-oblivious code authorship identification. In: Proceedings of the 2018 ACM SIGSAC Conference on Computer and Communications Security (2018)
2. Alam, S., Horspool, R.N., Traore, I., Sogukpinar, I.: A framework for metamorphic malware analysis and real-time detection. Comput. Secur. **48**, 212–233 (2015)
3. Arp, D., Spreitzenbarth, M., Hubner, M., Gascon, H., Rieck, K., Siemens, C.: Drebin: effective and explainable detection of android malware in your pocket. In: Ndss, vol. 14, pp. 23–26 (2014)
4. Carlini, N., Wagner, D.: Towards evaluating the robustness of neural networks. In: 2017 IEEE Symposium on Security and Privacy (SP), pp. 39–57. IEEE (2017)
5. Chouchane, M.R., Lakhotia, A.: Using engine signature to detect metamorphic malware. In: Proceedings of the 4th ACM Workshop on Recurring Malcode, pp. 73–78 (2006)
6. Christodorescu, M., Jha, S.: Testing malware detectors. ACM SIGSOFT Softw. Eng. Notes **29**(4), 34–44 (2004)
7. Christodorescu, M., Jha, S., Seshia, S.A., Song, D., Bryant, R.E.: Semantics-aware malware detection. In: 2005 IEEE Symposium on Security and Privacy (S&P 2005), pp. 32–46. IEEE (2005)
8. Clemens, J.: Automatic classification of object code using machine learning. Digit. Investig. **14**, S156–S162 (2015). https://doi.org/10.1016/j.diin.2015.05.007
9. Demetrio, L., Coull, S.E., Biggio, B., Lagorio, G., Armando, A., Roli, F.: Adversarial exemples: a survey and experimental evaluation of practical attacks on machine learning for windows malware detection. arXiv preprint arXiv:2008.07125 (2020)
10. Evtimov, I., et al.: Robust physical-world attacks on machine learning models. arXiv abs/1707.08945 (2017)
11. Eykholt, K., et al.: Robust physical-world attacks on deep learning visual classification. In: Proceedings of the IEEE Conference on Computer Vision and Pattern Recognition (CVPR), June 2018
12. Fu, J., Xue, J., Wang, Y., Liu, Z., Shan, C.: Malware visualization for fine-grained classification. IEEE Access **6**, 14510–14523 (2018)
13. Go, J.H., Jan, T., Mohanty, M., Patel, O.P., Puthal, D., Prasad, M.: Visualization approach for malware classification with ResNeXt. In: 2020 IEEE Congress on Evolutionary Computation (CEC), pp. 1–7. IEEE (2020)
14. Goodfellow, I.J., Shlens, J., Szegedy, C.: Explaining and harnessing adversarial examples. arXiv preprint arXiv:1412.6572 (2014)
15. Goodfellow, I.J., Shlens, J., Szegedy, C.: Explaining and harnessing adversarial examples. CoRR abs/1412.6572 (2015)
16. Grosse, K., Papernot, N., Manoharan, P., Backes, M., McDaniel, P.: Adversarial perturbations against deep neural networks for malware classification. arXiv preprint arXiv:1606.04435 (2016)
17. Han, K.S., Lim, J.H., Kang, B., Im, E.G.: Malware analysis using visualized images and entropy graphs. Int. J. Inf. Secur. **14**(1), 1–14 (2014). https://doi.org/10.1007/s10207-014-0242-0
18. Han, K., Lim, J.H., Im, E.G.: Malware analysis method using visualization of binary files. In: Proceedings of the 2013 Research in Adaptive and Convergent Systems, pp. 317–321 (2013)
19. AV-TEST Institute: Malware. https://www.av-test.org/en/statistics/malware/

20. Islam, R., Tian, R., Batten, L.M., Versteeg, S.: Classification of malware based on integrated static and dynamic features. J. Netw. Comput. Appl. **36**(2), 646–656 (2013)
21. Khormali, A., Abusnaina, A., Chen, S., Nyang, D., Mohaisen, A.: Copycat: practical adversarial attacks on visualization-based malware detection. arXiv preprint arXiv:1909.09735 (2019)
22. Kolosnjaji, B., et al.: Adversarial malware binaries: evading deep learning for malware detection in executables. In: 2018 26th European Signal Processing Conference (EUSIPCO), pp. 533–537. IEEE (2018)
23. Krcál, M., Svec, O., Bálek, M., Jasek, O.: Deep convolutional malware classifiers can learn from raw executables and labels only. In: ICLR (2018)
24. Krügel, C., Robertson, W.K., Valeur, F., Vigna, G.: Static disassembly of obfuscated binaries. In: USENIX Security Symposium (2004)
25. Liao, F., Liang, M., Dong, Y., Pang, T., Hu, X., Zhu, J.: Defense against adversarial attacks using high-level representation guided denoiser. In: Proceedings of the IEEE Conference on Computer Vision and Pattern Recognition (CVPR), June 2018
26. Liu, L., Wang, B., Yu, B., Zhong, Q.: Automatic malware classification and new malware detection using machine learning. Frontiers Inf. Technol. Electron. Eng. **18**(9), 1336–1347 (2017). https://doi.org/10.1631/FITEE.1601325
27. Liu, X., Zhang, J., Lin, Y., Li, H.: ATMPA: attacking machine learning-based malware visualization detection methods via adversarial examples. In: 2019 IEEE/ACM 27th International Symposium on Quality of Service (IWQoS), pp. 1–10. IEEE (2019)
28. Makandar, A., Patrot, A.: Malware class recognition using image processing techniques. In: 2017 International Conference on Data Management, Analytics and Innovation (ICDMAI), pp. 76–80 (2017)
29. Nataraj, L., Karthikeyan, S., Jacob, G., Manjunath, B.S.: Malware images: visualization and automatic classification. In: Proceedings of the 8th International Symposium on Visualization for Cyber Security. VizSec 2011. Association for Computing Machinery, New York (2011). https://doi.org/10.1145/2016904.2016908
30. Nataraj, L., Karthikeyan, S., Jacob, G., Manjunath, B.S.: Malware images: visualization and automatic classification. In: Proceedings of the 8th International Symposium on Visualization for Cyber Security, pp. 1–7 (2011)
31. Papernot, N., McDaniel, P., Wu, X., Jha, S., Swami, A.: Distillation as a defense to adversarial perturbations against deep neural networks. In: 2016 IEEE Symposium on Security and Privacy (SP), pp. 582–597. IEEE (2016)
32. Pierazzi, F., Pendlebury, F., Cortellazzi, J., Cavallaro, L.: Intriguing properties of adversarial ml attacks in the problem space. In: 2020 IEEE Symposium on Security and Privacy (SP), pp. 1332–1349. IEEE (2020)
33. Rad, B.B., Masrom, M., Ibrahim, S.: Camouflage in malware: from encryption to metamorphism. Int. J. Comput. Sci. Netw. Secur. **12**(8), 74–83 (2012)
34. Raff, E., Barker, J., Sylvester, J., Brandon, R., Catanzaro, B., Nicholas, C.: Malware detection by eating a whole EXE. In: AAAI Workshops (2018)
35. Shaid, S.Z.M., Maarof, M.A.: Malware behavior image for malware variant identification. In: 2014 International Symposium on Biometrics and Security Technologies (ISBAST), pp. 238–243. IEEE (2014)
36. Sharif, M., Bhagavatula, S., Bauer, L., Reiter, M.: Accessorize to a crime: real and stealthy attacks on state-of-the-art face recognition. In: Proceedings of the 2016 ACM SIGSAC Conference on Computer and Communications Security (2016)

37. Sharif, M., Lucas, K., Bauer, L., Reiter, M.K., Shintre, S.: Optimization-guided binary diversification to mislead neural networks for malware detection. arXiv preprint arXiv:1912.09064 (2019)
38. Tong, L., Li, B., Hajaj, C., Xiao, C., Zhang, N., Vorobeychik, Y.: Improving robustness of {ML} classifiers against realizable evasion attacks using conserved features. In: 28th {USENIX} Security Symposium ({USENIX} Security 19), pp. 285–302 (2019)
39. Xiaofang, B., Li, C., Weihua, H., Qu, W.: Malware variant detection using similarity search over content fingerprint. In: The 26th Chinese Control and Decision Conference (2014 CCDC), pp. 5334–5339. IEEE (2014)
40. Zhang, G., Yan, C., Ji, X., Zhang, T., Zhang, T., Xu, W.: DolphinAttack: inaudible voice commands. In: Proceedings of the 2017 ACM SIGSAC Conference on Computer and Communications Security (2017)
41. Zhang, Q., Reeves, D.S.: MetaAware: identifying metamorphic malware. In: Twenty-Third Annual Computer Security Applications Conference (ACSAC 2007), pp. 411–420. IEEE (2007)

A Survey on Common Threats in npm and PyPi Registries

Berkay Kaplan[✉][ID] and Jingyu Qian[ID]

University of Illinois at Urbana-Champaign, Urbana, IL 61801-2302, USA
{berkayk2,jingyuq2}@illinois.edu

Abstract. Software engineers regularly use JavaScript and Python for both front-end and back-end automation tasks. On top of JavaScript and Python, there are several frameworks to facilitate automation tasks further. Some of these frameworks are Node Manager Package (npm) and Python Package Index (PyPi), which are open source (OS) package libraries. The public registries npm and PyPi use to host packages allow any user with a verified email to publish code. The lack of a comprehensive scanning tool when publishing to the registry creates security concerns. Users can report malicious code on the registry; however, attackers can still cause damage until they remove their tool from the platform. Furthermore, several packages depend on each other, making them more vulnerable to a bad package in the dependency tree. The heavy code reuse creates security artifacts developers have to consider, such as the package reach. This project will illustrate a high-level overview of common risks associated with OS registries and the package dependency structure. There are several attack types, such as typosquatting and combosquatting, in the OS package registries. Outdated packages pose a security risk, and we will examine the extent of technical lag present in the npm environment. In this paper, our main contribution consists of a survey of common threats in OS registries. Afterward, we will offer countermeasures to mitigate the risks presented. These remedies will heavily focus on the applications of Machine Learning (ML) to detect suspicious activities. To the best of our knowledge, the ML-focused countermeasures are the first proposed possible solutions to the security problems listed. In addition, this project is the first survey of threats in npm and PyPi, although several studies focus on a subset of threats.

Keywords: Open-source · PyPi · npm · Machine learning · Malware detection

1 Introduction

The OS movement allows any developer to read or attempt to contribute to the source code hosted in a publicly accessible location, such as GitHub. This trend may have several advantages, such as code reviewers catching software bugs promptly. Furthermore, the creativity behind a software's design is no longer

© Springer Nature Switzerland AG 2021
G. Wang et al. (Eds.): MLHat 2021, CCIS 1482, pp. 132–156, 2021.
https://doi.org/10.1007/978-3-030-87839-9_6

limited to a team of developers as anyone is free to propose their ideas. This advantage allows additional features or functionalities to be implemented, thus possibly making the project the market leader in its area.

Although several advantages are present with the OS movement, there are also disadvantages. One possible disadvantage is that certain individuals in public may have malicious intentions, and the publicly accessible code may raise security concerns. A team of founders of the OS platform may review every pull request to ensure the benign intentions of the commits. However, some environments are known to lack control measures. Some of these environments may include public registries, which are public repositories anyone can upload their packages to for others to use. These repositories intend to serve as collections of libraries to facilitate development tasks with readily available packages. Some public registries do not have any control measurements besides reporting the malicious code once detected by a developer. This issue gives the attacker a certain lifespan to perform malicious activities.

PyPi and npm are two popular public registries. These environments require developers to verify their email addresses without further control, making it much easier for attackers to create malicious accounts. These accounts can publish any package in their public registry, amplifying an attack's impact naturally. Furthermore, attackers can take advantage of several other techniques that benefit from human errors. Some of these techniques are typosquatting and combosquatting. When an attacker intentionally publishes a package in the public registry with a similar name to popular packages, these techniques benefit from a potential typo during the manual installation of dependencies [35]. Typosquatting takes advantage of typos during the installation of packages, while combosquatting hopes to exploit a mistake of the order in the package name, where the name consists of multiple nouns, such as "python-nmap" typed as "nmap-python" [35]. The ultimate goal in both attacks is to get the developer to install the malicious package.

Another popular attack surface for these registries is the interdependent nature of its packaging environment. A handful of accounts maintain numerous packages. These packages are dependent on several other libraries. Compromising these accounts would give an attacker the ability to modify the code at will. Furthermore, if there is a vulnerability in one of the popular packages, then the issue will propagate through all the other libraries that depend on it. This concept is also known as software supply-chain security, which refers to outsiders accessing the final project [11,21]. Software supply-chain security is another research field but seems to be heavily related to OS registry security. There has already been a study on software supply-chain and typosquatting attacks on public registries that includes npm and PyPi; however, the study did not cover other common dangers in the OS registries such as trivial packages [10].

Another attack surface in the OS settings is the technical lag, which is the time or version difference between the dependency used in production and the most recent version available for installation [39]. For instance, when package A depends on several other libraries in production that are not up to their latest versions, known vulnerabilities may compromise package A as well through these dependencies. npm does not seem to help developers keep up with the latest versions of their libraries, as package.json requires developers to use a specific version or range of versions of a package instead of automatically choosing the most recent version [41].

We are the first to present a systematic survey of public registries' common vulnerabilities. In this paper, we will be exploring the impact of each vulnerability. We will be suggesting untested risk mitigation strategies we compiled from the literature to the administrators of the mentioned OS registries. We specifically incorporated ML techniques into our risk mitigation strategies to make our solutions more adaptable to different types of future attacks. We made this assumption as the features of ML-based tools can be changed to modify the program's purpose.

To the best of our knowledge, our work is the first in the field to offer countermeasures against the security problems in npm and PyPi environments. We have not conducted physical experiments or surveyed developers, as we merely focused on presenting the state of the field. We outlined our main contributions to the field:

- Compilation of all the threats and vulnerabilities from the literature that exists in public registries.
- Compilation of work that explains the prevalence of the presented risks in popular OS environments.
- Countermeasure suggestions against the mentioned attack surfaces.

Similar work has been published previously, such as the study from Vaidya et al. in 2019. The paper indeed built a systematic threat model in OS environments and provided threat classifications. Nevertheless, we determined that this study did not provide sufficient coverage as certain issues, such as trivial packages, were not discussed.

The OS movement has brought several advantages, and we would like to support the safety of its members. Supporting the public registry safety will also benefit the OS community as it would draw some companies, which could not enter the environment due to their high-security requirements. This paper aims to disclose the security issues in public registries to contribute to the OS community. This project's scope does not extend to vulnerabilities in native code or compiler as we focus more on software reuse.

2   Background

It would be beneficial to explain a couple of frequently used terms in this paper that relate to the interdependent structure of libraries.

2.1   Dependencies and Dependency Trees

Dependencies, packages, or libraries are fairly similar terms in the context of our survey. Dependencies are readily available code for developers to use in their projects to accelerate the speed of the implementation phase. Private companies may create these dependencies, or they may even be OS. The implementation in these libraries is usually not individually scrutinized for efficiency purposes in a project's life cycle.

A dependency tree visualizes the direct relationship between libraries. The simplest tree structure would be that package A depends on package B, meaning that package A uses components, functions, or code snippets from package B. These relationships can extend to several packages vertically, introducing more attack surfaces for a specific dependency.

2.2   Package Manager

Package managers allow developers to manage their dependencies in a central document, for example, a JSON file. The managers facilitate the installation and sharing of dependencies as the developer no longer has to visit the library's official website to download the package into his local repository [9]. The central document has a list of the dependencies. It is already connected to a registry that contains all the libraries needed via software. These managers are also configurable to connect to private registries, where all the libraries are internal to a company. Recent studies have shown that package managers have been abused to distribute malware, thus making them a trade-off between convenience and security [9,33].

2.3   Software Supply-Chain Attacks

The goal of supply-chain attacks is to inject malicious code into software or library externally. These attacks modify the targeted program so that it is still validly signed by its owner. The attacker simply injects his code into one of the software's dependencies [21]. Theoretically, an attacker can do this injection in any given node in the software's dependency tree. The ultimate goal is to alter the behavior of the root or a specific node, which is the targeted software product, using a child node, which in this case is a dependency node in the tree [21].

2.4   Typosquatting and Combosquatting

Typosquatting is an attack methodology where an attacker intentionally publishes a package with a typo in its name, making the name relatively close to a popular package's name. The attacker hopes that a victim would make a typo while downloading dependencies. Thus, the victim would download the malicious package instead of the intended one. One example of this type of attack is the popular *lodash* package incident, where it had a typosquatting package named *loadsh*. In the English language, a developer might likely download the *loadsh* package after making a typo.

Combosquatting is similar to typosquatting; however, some packages consist of multiple words in their name. Sometimes, it would be easy to forget the order of the words, and the attacker intentionally creates a package that contains the same words as the targeted package, but the order of the words will vary. The goal is similar to typosquatting, benefiting from an innocent mistake of a developer. A researcher uploaded typosquatting packages to different repositories, including PyPi, and the package received around 45K downloads over several months [35].

2.5   Machine Learning

Learning is considered the hallmark of human intelligence, and researchers have worked on making machines intelligent, giving birth to ML [36]. It is the study of methods to allow computers to gain new knowledge and constantly improve themselves [36]. The definition can also be the process that allows computers to convert data to knowledge through certain techniques [36]. The applications of the ML techniques have been quite popular in recent years, and one of these applications is the area of information security. We will incorporate some ML techniques into our proposed security remedies to provide a more reliable method of identifying and preventing malicious actors.

3   Motivation

Although several studies exist in the field of OS registry and software supply chain security, there has not been a known source to publish a survey-structured paper to construct a centralized information bank on the types of dangers in OS registries. Furthermore, we have not found a project to offer mitigation strategies against these attacks. The feasibility of our solutions may be unclear, as we have not tested any of them. However, we hope they would serve as a first step to develop more systematic countermeasures.

OS registry security is a relatively generic field that tackles certain security issues in OS environments. It quite surprised us that a survey does not yet exist on the security of the public library registries. Therefore, to contribute to the field and give a glimpse of the area's current state, this work has been thought to be useful. We hope to enhance the information security field by offering risk mitigation strategies and serving as a first step to build a novel chapter in its curriculum. We expect to have several corrections done on this draft of a chapter. Still, our ultimate motivation is to foster interest in the field and raise awareness against these issues as we also support the OS community.

4   General Overview of Vulnerabilities in npm and PyPi

Although several other package managers and OS environments exist, we decided to focus on PyPi and npm as they contain the most libraries in their registries. Similar ecosystems and their size comparisons are illustrated in Fig. 1 [33].

To understand the scope of the security issue and get an estimation of the number of vulnerable packets, we scanned research papers. We found that one of

Fig. 1. A size comparison in package management ecosystems [33]

the projects in our references compiled a summary of 700 security reports gathered from Snyk.io published before 2017-11-09 in Fig. 2 [7]. However, researchers removed certain issues, such as typosquatting, as they do not introduce vulnerabilities in existing packages [7]. Among the 610K packages, 133602 packages directly depend on vulnerable packages, and 52% of these packages, which is 72470 in total, have at least one release that relies on a vulnerable version [7].

610,097	npm packages
4,202,099	releases of npm packages
20,240,402	runtime dependencies
399	security vulnerability reports
269	packages affected by the vulnerability
14,931	releases of such vulnerable packages
6,752	releases affected by the vulnerability
133,602	packages depending on a vulnerable package
72,470	dependent packages affected by the vulnerability

Fig. 2. A summary of the npm dataset [7]

The median number of releases affected by a high-severity vulnerable package is 26 in the npm environment [7]. The encouragement from the npm and PyPi environment to reuse code seems to accelerate this issue [33].

4.1   Direct, Indirect Dependencies, and Heavy Code Reuse

As stated before, code reuse is common in the npm environment. Projects are either directly or indirectly dependent on each other, meaning that project A that uses code from project B has a direct dependency relationship as A directly depends on B. If B depends on project C, then A has an indirect dependency on C, increasing the attack surface. Indirect dependencies are also referred to as "transitive dependencies" as some of the work in the literature uses that term to build their graphs, such as Fig. 3 that indicates the average number of dependencies projects use over time.

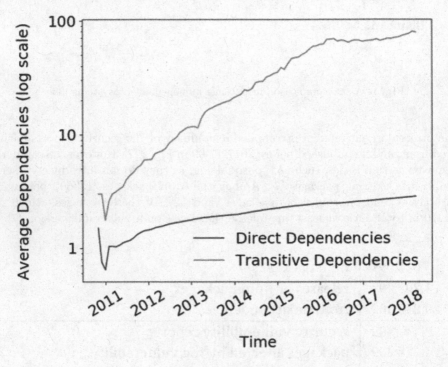

Fig. 3. The growth of both types of dependencies over time in logarithmic scale [41]

Figure 3 implies that a developer implicitly trusts around 80 other packages by using a library [41]. This trust also means that a hacker's attack surface includes the accounts that maintain the indirectly dependent packages.

Another term based on transitive dependency is the "Package Reach," which refers to the number of transitive dependencies a specific library has [41]. It has been stated that some popular packages have a package reach of over 100000, drastically increasing the attack surface, and this trend has been observed to be increasing [41]. The package reach also increases the size of the dependency graph. If one of the dependent packages is compromised by either a squatting attack or a stolen credentials of a maintainer account, it would be reasonable to assume that a security risk would arise. The average package reach for a single given library seems to be in an increasing trend for the last couple of years, according to Fig. 4.

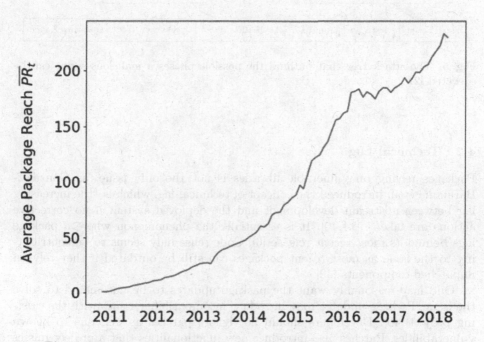

Fig. 4. The average package reach overtime in the npm environment [41]

Once the attacker injects his code into a dependency graph, he can execute it in various phases. The attack tree that outlines wherein an attacker can execute the malicious script is also illustrated in Fig. 5.

Zimmerman's 2019 USENIX paper seems to be a leading study examining the interdependent structure of packages and their trends in the npm environment. However, it has also been determined that most papers in this field seem to be less than five years. This finding implies that security experts recently started to investigate the interdependent packaging structure of OS environments. Software supply-chain attacks, on the other hand, seemed to be a more mature field.

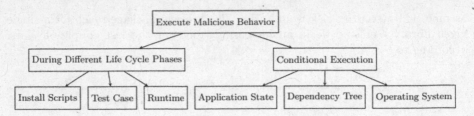

Fig. 5. The attack tree that outlines the possible phases a malicious script can be executed [21].

4.2   Technical Lag

Packages relying on vulnerable libraries is not the only issue, as Gonzalez-Barahona et al. introduced the concept of technical lag, which is "the increasing lag between upstream development and the deployed system if no corrective actions are taken" [14, 40]. It is essentially the phenomenon where a package lags behind its most recent release [6]. Code reuse only seems to be contributing to the issue as most recent packages can still be outdated if they rely on unpatched components [40].

One may reasonably want the package updates to be automated to solve the issue. Package updates may introduce incompatibility issues with the existing project [40]. Developers should not ignore patches as well due to known vulnerabilities. Patches also introduce new functionalities that a project misses by avoiding incompatibility risks, causing a concept named "technical debt" to arise [14]. However, the issue mainly seems to be a trade-off between security and functionality.

The perspective against technical lag also seems to vary as the literature contains data implying that clients can be safe from technical lag. It has been concluded that one in four dependencies and two in five releases suffer technical lag in the npm environment [6]. After 2015, it has also been observed that the average duration of technical lag in npm was 7 to 9 months, meaning that a specific dependency would be updated 7 to 9 months after a library's new release [6].

Furthermore, benign users might not be the ones to discover a vulnerability first. Most vulnerabilities take a long time to be discovered, especially low-severity ones [7]. Although project contributors fix most vulnerabilities discovered in the source code in a short amount of time, a non-negligent portion takes longer [7].

To counter the dangers of the technical lag argument, we found one report stating that 73.3% of outdated clients were safe from threats as these clients did not use the specific vulnerable code [38]. A vulnerability tends to be in a specific function, and not every client will have to use that particular function. However, the risk of a client using the vulnerable function persists. Developers who realize the functions they are using are not particularly affected by the stated issue take longer to update their dependencies, increasing the technical lag [38].

In addition, One may reasonably conclude that the older the package is, the less vulnerability it will have. Thus, a possible solution may be to use old packages. However, a study in the literature has found that this claim is wrong as they found that most vulnerabilities are in packages older than 28 months [7]. Older packages tend to have more severe vulnerabilities [7]. We could not establish a relationship between security and the package's age in this study.

4.3   Squatting Attacks

We found that attackers mainly use two methods to spread malicious code in these environments: stealing the credentials of accounts that maintain certain packages to inject their code into the project and tricking users into downloading their packages by methods such as squatting attacks [35]. The term squatting attack is an umbrella concept used in this paper to represent typosquatting and combosquatting. We will discuss other types of attack methods against the public registries in this paper, but we focus on these two scenarios in this section. In the first scenario, the "ssh-decorate" package was affected as the attackers took control of the maintainer account and injected code into the package. The injected code would send users' ssh credentials to an attacker-controlled remote server [35]. Some popular maintainers can contribute to hundreds of packages, making their accounts critical to a potential compromise [41].

In the latter scenario, attackers can fork an existing package and modify the *setup.py* file to download the malicious dependencies in PyPi [35]. In addition, an attacker can even use an existing package name after a popular package has been withdrawn by the maintainer from the registry, introducing further attack surfaces [21]. One such attack had happened in the past with the go-bindata package when its owner deleted the unmaintained package vaidya2019security. Fortunately, the attack was discovered on the same day [33]. Squatting attacks are a problem in the OS registry environment because the repositories allow users to upload code as users are given equal trust initially [30].

Squatting attacks are both dangerous for the developer, stakeholders, and the client as a Trojan virus may be embedded that activates when the project is run [30]. Certain code runs on users' machines when npm or PyPi installs packages on the local repository. Thus a client does not even have to use the dependency to be attacked. Some packages are even known to open reverse shells with user privileges using the installation scripts of the packages [30]. Manual squatting detection proves to be challenging as some of the name changes are quite hard to notice, as illustrated in Table 1 that shows some past attacks [30].

The dangers of a typosquatting attack come from the fact that the client is not affected by the malicious activity only but all the other libraries that depend on it, meaning that a developer does not have to make a typo mistake [30]. Another challenge in catching these attacks is that not all typosquatting packages are malicious, as the *loadsh* example did not have any harmful content [30]. However, it was a copy of an older version of the original *lodash*, meaning

Table 1. Some of the samples of squatting attacks selected from the literature in the PyPi environment [35]

Malicious Package	Legitimate package	Names change
virtualnv	virtualenv	Delete 'e'
mumpy	numpy	Substitute 'n' by 'm'
django-server	django-server-guardian-api	Delete "-guardian-api"
urlib3	urllib3	Delete 'l'
python-mysql	MySQL-python	Swap "python" and "mysql"
python-openssl	openssl-python	Swap "openssl" and "python"

that public unpatched vulnerabilities still may be a security concern, as it has been reported that 63 other packages depended on *loadsh* [30].

4.4   Maintainers and Collaborators

There is a distinction between maintainers and collaborators in OS environments. Maintainers review and approve contributions while contributors propose code changes [21]. Both of these accounts are critical for a secure project. An attacker's goal is to inject malicious code into the dependency tree to affect the targeted software. An attacker can reach that goal by either creating a new package via squatting or using the existing packages that already have users [21]. If a maintainer account is compromised, an attacker can easily inject his code. Another possible scenario is that an attacker can pose as a benign contributor and make pull requests with a seemingly useful feature, such as the incident of the unmaintained "event-stream" package that will be mentioned in the following sections [21,33]. Embedded inside the feature can be malicious content that may be hard to notice, and considering that a human is reviewing the request, human error can allow the request to be approved. Alternatively, a maintainer account with weak credentials can always be compromised to approve malicious pull requests, or attackers can use social engineering strategies to perform the injection.

The trend presented in the transitive dependency structure is not the only alarming graph. The rise of npm would be from new developers joining the system mainly; however, the package per maintainer has also increased as current members publish new projects [41]. Each account is becoming responsible for more packages, making it more critical to secure accounts. The average package count per maintainer account seems to be 2.5 in 2012, ascending to 3.5 in 2013 and almost 4.5 in 2018 [41]. This increase happened for both average and especially for influential maintainers in npm, as the case for some selected popular maintainer usernames can be observed in Fig. 6 [41].

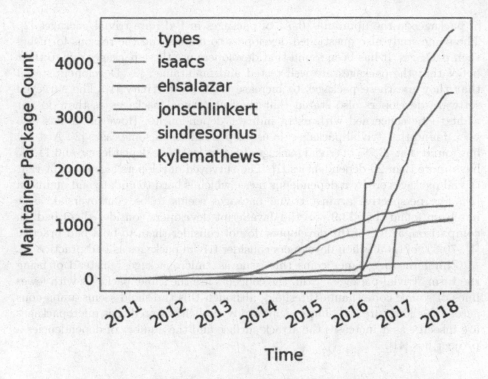

Fig. 6. Package count per popular username in npm [41]

4.5   Trivial Packages or Micropackages

Trivial packages are small libraries, which a study in the literature found to be less than 35 lines of code, according to 79% of their participants who attempted to classify libraries as trivial [2]. Developers have considered trivial packages good until recent trends pushed code reuse to an extreme [2]. The breakdown of popular web services, such as Facebook, Airbnb, and Netflix in 2016 from a Node.js 11-line trivial package named left-pad, only made the concept be questioned further [1,5].

The incident has been referred to as the case that "almost broke the Internet," which led to many discussions over code reuse sparked by David Haney's blog post "Have We Forgotten How to Program?" [1]. Node.js used to allow developers to unpublish projects that lead to the left-pad incident emerge [1]. Although the origin of the incident is not related to a vulnerability in the package itself, it has raised awareness about trivial packages [1]. Numerous developers agreed with Haney's opinion that developers should implement trivial tasks themselves rather than adding an extra dependency to the project [1]. Since then, people have been working to investigate this issue.

A study has researched the developers' perspectives of trivial packages or micropackages, and the interviewed developers stated that their definition of micropackage is the same across PyPi and npm [2]. It has been found that 16%

of packages in the npm and 10.5% of packages in PyPi are trivial packages [2]. The same study also questioned developers to understand the reasons for using such packages. It has been found that developers use these packages due to their belief that the packages are well tested and maintained [2]. Developers stated that they use these packages to increase their productivity [1]. The surveyed software developers also stated that they use trivial packages as they do not want to be concerned with extra indirect dependencies. However, it has also been found that trivial packages do use dependencies in some cases [2]. A study has found that 43.7% of trivial packages have at least one dependency, and 11.5% have more than 20 dependencies [1]. The surveyed developers also thought that trivial packages create a dependency mess, which is hard to update and maintain [1]. The perspective against trivial packages seems to be controversial, as it has been found that 23.9% of the JavaScript developers consider them bad. In comparison, 57.9% of the developers do not consider them to be a bad practice [1]. 70.3% of the Python developers consider trivial packages as bad practice [2].

Zimmermann's paper coins this issue as "micropackages" instead of using the term "trivial package." Still, the concepts are the same: packages with fewer lines of source code than a threshold, although this threshold seems ambiguous across the literature [41]. The specific study explicitly stated that micropackages are insecure as it increases the attack surface and the number of dependencies a project has [41].

4.6 PyPi Overview

PyPi has limited automated review tools for the uploading process as the npm environment does, making it vulnerable to different kinds of attacks, such as squatting [35]. Furthermore, the moderator and administrator team, who has permission to remove packages from the registry, seems to be less than ten people, limiting the maximum number of code reviews they can conduct [35]. Considering the 400K package owner, each administrator seems to be responsible for 40K people, assuming every administrator performs code review for malicious content, thus providing a lower bound for the number of package owners per moderator ratio. PyPi allows end-users to report malicious packages. Nevertheless, considering each moderator being responsible for at least 40K developers, it would be only reasonable to question the efficiency of the code reviews. When users download packages using the *pip*, there is no available system that reviews the code to determine its safety aside from a user's antivirus. So, we can outline the process of publishing packages with Fig. 7 that illustrates a high-level view of the schematics of the PyPi ecosystem.

Spreading malicious code in the wild PyPi is fairly similar to the process we illustrated in the npm environment. Mainly, an attacker can either steal the credentials of an existing account to exploit the current reputation of the project or create a new package by forking the targeted package and modifying the content, or simply creating a brand new package [35]. The latter method can still trick users into downloading the attacker's library by a squatting method.

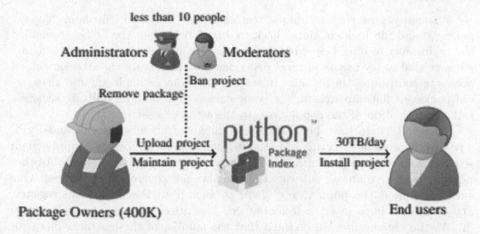

Fig. 7. An overview of the roles in the PyPi ecosystem [35]

4.7   Noteworthy Incidents

Although we described the attack types, we have not given detailed concrete examples until this point. Thus, it would be fruitful for the reader to illustrate these concepts with real-life scenarios.

In July 2018, an attacker stole the credentials of a developer account of a package. The attacker published a malicious version of the *eslint-npm* package that would, during installation, send users' npmjs.org credentials from the .npmrc file to an attacker-controlled server[33]. Although the attack was detected the same day, it has been estimated that 4,500 accounts may have been compromised [33]. This incident clearly shows the importance of the period until an attacker is detected.

A known typosquatting attack happened in July 2017 when a user named "HackTask" uploaded 40 malicious packages to the npm registry with similar names to popular packages. The packages had payloads in their installation script that would ex-filtrate local environment variables, storing sensitive authentication tokens, to the attacker's server [33]. The attack was discovered after 12 days, and it has been estimated that 50 users downloaded the attacker's packages [33].

Another typosquatting attack happened in the PyPi public registry, where a malicious package, named *jeIlyfish*, hoped to imitate the popular package *jellyfish* by changing the first L of jellyfish to an I, and it stole ssh and GPG keys [34]. The malicious package existed in the repository for a year until it was discovered, thus implying that squatting attacks do not necessarily get discovered in a short period [34]. One study, which also expressed that vulnerability detection and prevention are insufficient in PyPi, found that the median time a vulnerability is hidden in the PyPi environment is three years regardless of their severity[4]. Once the vulnerability is discovered, on median, all packages take four months to have their vulnerability patched, giving plenty of time for an attacker to exploit a zero-day vulnerability [4].

A social engineering technique was also used to attack the npm "copay" package, and the incident dates back to July 2018, when the attacker emailed the maintainer to offer help with the popular but unmaintained "event-stream" package that copay has an indirect dependency on [33]. Once the attacker gained access to contribute, the attacker injected a malicious dependency into an external package, "flatmap-stream," of event-stream that would ex-filtrate sensitive wallet information of the copay users to the attacker's server [33].

All in all, there have been several examples of attacks in npm and PyPi. The interdependent structure of both environments raises software supply-chain security as a concern. It is not only npm and PyPi that gets targeted but other popular package manager software that we have not covered in this survey. One such example is the popular *rest-client package* from the RubyGems registry. attackers used it to insert a Remote-Code-Execution backdoor on web servers [9]. A study in the literature stated that the number of these attacks increases [4,34]. Therefore, we expect to see additional incidents in the field if admins do not implement countermeasures.

5   Discussion

This research focused on presenting the costs of using OS registries to benefit from the readily available libraries other developers have created. These costs mainly originate from security concerns that may damage institutions' IT infrastructure. Evidence has been gathered from the literature to study the main threats in the npm environment comprehensively. No further experiments or studies have been done in this work, as we solely focused on compiling recent work in OS registry security. We determined that the OS environment provides new attack surfaces to inject code into unsuspecting victims' machines. An attacker injecting a script into a victim's computer can have devastating effects, considering the Turing-completeness of these languages.

5.1   Suggested Countermeasures

We have developed countermeasures based on the ideas we gained from our literature search. We specifically attempted to incorporate ML techniques into our approach. We took this step as the offered solutions will only need minor modifications to features to cover new undiscovered future attacks. This advantage of ML will make the presented techniques more adaptable. Furthermore, although signature-based malware detection is the most extended technique in commercial antivirus, they tend to fail as newer types of malware emerge [24,26]. Artificial intelligence provides a competitive advantage over next-generation malware [26].

Therefore, when designing our countermeasures for public registries, we incorporated ML techniques. We have not tested the solutions yet, so the feasibility of using them may be in question. Another option can be contacting the admin teams of the public registries to gain their inputs. However, we hope that this work would act as a first step to spark interest in this field and capture the attention of the admin teams of OS registries.

Malicious Package Detection. Some sources state that current tools for identifying malicious payloads are resource-demanding. However, there are lightweight tools in the literature to mitigate the risk of malware residing in the registries. These tools can be automated scanning programs to analyze the dependency tree for squatting packages [34]. The automation of reviewing published code may prove to be challenging. Nevertheless, allowing any developer to publish code into the public registry, where millions of developers have access, gives much power to any stranger.

Since the source code for every project on an OS registry is public, any user can review the code for each library. This fact is advantageous as the dependency tree and the source code of the project is visible. One antivirus designed for APIs can review packages on the registry to eliminate the time needed for the public to uncover malicious software. A member of the admin team or a volunteer can download batches of packages from the public registry to his local repository and use the antivirus to ensure the safety of each library. However, we have not found an antivirus software specifically designed to scan APIs on the internet. Antiviruses may use features that work on software for end-users. The same features may not prove efficient for libraries designed for implementing programs.

For this purpose, we will explore antivirus software that may be useful specifically for APIs. We will focus on analyzing the binary and the behavior of the library instead of its source code. We decided to cover a broader range of use cases, as we believe analyzing the source code is less challenging than an executable binary. Thus, our project can also cover library binaries for malware detection.

There has been work in the field to efficiently automate the detection of malicious code injection in the distributed artifacts of packages, and the admins may attempt to implement some of these novel tools [34]. One such tool, named *Buildwatch*, analyzes the third-party dependencies by using the simple assumption that malicious packages introduce more artifacts during installation than benign libraries [22]. This hypothesis has been formulated and tested in the *Buildwatch* study [22]. Admins can also modify *Buildwatch* to detect squatting packages in the dependency tree with the techniques we will discuss in the upcoming sections.

One approach in detecting suspicious packages comes from a study that uses anomaly detection to identify suspicious software [13]. Their approach first extracts several features from the version when a package gets updated. It then performs k-means clustering to detect outliers for further review. The researchers kept the number of features low to ensure the clustering algorithm takes limited computation power, thus making it lightweight [13]. Some extracted features are the ability to send/receive HTTP requests, create/read/write to file systems, or open/listen/write to sockets [13]. This tool specifically uses anomaly detection, a branch of ML, to tackle the issue of detecting suspicious updates. We believe this tool is very suitable for analyzing third-party libraries.

As image classification becomes more accurate due to the development of robust classifiers such as convolutional neural networks (i.e., CNN), researchers in the field proposed visualization-based techniques to detect malicious packages. This technique would input the executable of the library to visualize its binary for processing. An advantage of binary visualization is that it does not need to disassemble the package to perform static analysis if the source code is not available. The program can directly examine the package binary.

Nataraj et al. [19] proposed a method to convert the malware binary into a gray-scale image by reading the binary as vectors of 8 bits and organizing them into a 2D array. They extracted texture features from the image representation of the malware and performed K-nearest neighbor (i.e., KNN) with Euclidean distance to classify the malware.

Ni et al. [20] converted malware opcode sequences into gray-scale malware fingerprinting images with SimHash encoding and bilinear interpolation and used CNN to train the malware classifier. Their evaluation results revealed that their classifier outperformed support vector machine (SVM) and KNN on the features from 3-grams of the opcode, and another classifier using KNN on GIST features. The classifier reaches around 98% accuracy and only takes 1.41s to recognize a new sample.

Besides the visualization of the binaries of libraries, which is similar to static analysis, there are other ML-based techniques. The ML-based malware detection field seems well-established as several survey papers exist, such as studies from Gandotra et al. and Ucci et al. [12,31]. They compiled various tools from the literature, and Gandotra et al. listed their limitations [12]. These tools are based on static analysis, in which the code is not executed, and dynamic analysis, which focuses on the run-time behavior [12]. One such tool even uses data mining techniques to detect malicious executables that rely on specific features from the binary. [25]. These features are the Portable Executable, strings, and byte sequences [25]. The Naive Bayes model takes in these features and reaches an accuracy rate of 97.11% in the study [12,25]. These surveys contain several examples of malware tools that can be adapted for APIs.

In 2014, Gandotra et al. stated that many researchers lean towards dynamic analysis instead of static [12]. The survey gave examples of dynamic analysis techniques for malware detection [12]. Some examples include the work of Rieck et al. that monitors the behavior of malware on a sandbox environment [12,23]. The observed behavior of the program is embedded into a vector space [23]. They use clustering to identify similar malware behaviors to classify each executable [23].

To conclude the antivirus examples summary for APIs, we believe that the volunteer developers recruited via an advertisement banner on the official npmjs.com website can implement the mentioned tools and ML techniques in exchange of the trust score incentive in the following sections. Afterward, the admin team can design a protocol where a volunteer user can install batches of libraries into his local repository to allow the mentioned programs to review their safety. Developers can even automate this protocol from a web server to prevent the loss of precious human labor on repeated tasks.

If our methods are not feasible for public registries, novel tools still exist in the literature accessed from the Google Scholar platform. We could not find existing antivirus explicitly designed for libraries, but already-tested software would be a better option if admins can access it. Admins can decide to move forward with a novel tool originated from a paper, and reading academic papers can indeed be a daunting task as it is for us. However, most papers have correspondent authors that can clarify ambiguous implementation details. Some papers even have their implementations on public repositories, which will be very helpful to the team of implementers.

Reasonable Constraints, Decentralization, and Trust Score. The founding committee of an OS registry might not operate to pursue profit, and they may be short on budget to hire additional developers to implement these tools or reviewers to find malicious packages manually. However, to solve this problem, the admin teams can decentralize the manual code review process to people willing to contribute to the platform's security. Contributors can opt-in to review code and cast votes. After a new library is reviewed and voted to be published by a certain number of developers, it may be available to install on the public registry.

To encourage members of the OS registries to become reviewers, admin teams of the registries can give specific incentives for the time a reviewer invests. These perks, such as early access or discounts to products, can be supplied by companies that sponsor the npm or PyPi environment. As reviewers cast their votes to pass or reject a commit, they can earn points to unlock some advertised perks. The point system can be gamified to build public rankings and hierarchic levels. Each level would give contributors different perks, such as a free course on a partnered educational platform.

To implement the final stage of the voting-based solution, one has to take measures against the possibility of reviewers abusing the scoring system by casting votes without properly reviewing a commit. We believe there are already known solutions for this issue that does not need to be mentioned in this work. Some of these control measures can include comparing the time it takes for each reviewer to cast a vote to ensure each reviewer invests a proper amount of time. Simply observing the number of times a reviewer is an outlier based on the majority of the votes cast for a certain decision can also be observed. These solutions are merely to preserve the quality of the reviewer committees.

The scoring system can be implemented to give more privileges to users gradually besides the mentioned perks. Users can earn trust points as they actively review the contributions to projects, and users may start to publish their packages after their score passes a threshold. Thus, registries can give power to partially trusted users rather than anyone with an internet connection. We will be referencing this trust score in some of the following sections.

Securing Critical Accounts. We believe the admin teams should implement a mandatory feature in the registries. A user who maintains a popular library or a popular library that uses the user's package should automatically have multi-factor authentication (MFA) to mitigate the risk of compromised accounts. MFA should be mandatory in highly trusted accounts that pass a preset score threshold as well. Any abnormal behavior, such as login from different machines, should be recorded and processed accordingly to prevent more sophisticated attacks. This process can simply be a secret question or a one-time code.

Another possibility is integrating code authorship identification systems into OS registries, where the program would get triggered whenever a code is published from a known author. Known hackers can be profiled from their public source code to create a database of dangerous code styles that may indicate compromised accounts. This practice is a research field in itself, and it would require high investment to build such an infrastructure. However, there are already systems built to tackle this issue to an extent. One such system is demonstrated in a study that demonstrates a Deep Learning-based Code Authorship Identification System (DL-CAIS), which is language oblivious and resilient against code obfuscation [3]. DL-CAIS uses GitHub repositories as its dataset and reaches an accuracy of 96% [3]. It first preprocesses the input code to build representation models and feeds those models into the deep learning architecture to find more distinctive features [3]. The architecture to be trained consists of several RNN layers and fully connected layers [3]. After the training, researchers use the resulting representation models to construct a random forest classifier [3].

Though limiting ourselves to only known hackers may be an insufficient approach. However, npm and PyPi have the source code of each project, proving to be very advantageous to profile developers. npm or PyPi can use code authorship detection tools to profile each of its developers. Each constructed profile can be associated with a collaborator or maintainer account. If the uploaded code profile in one commit does not match the code profile of the account, the admin can place a temporary hold on the account for further verification. One may still be concerned about the privacy aspect of this approach, but the feature can be optional for non-critical accounts. An automated script can reward users with an extra trust score for opting into the program to provide incentives for this practice. However, we believe critical accounts should be required to participate, as the cost of an account compromise would be significantly higher in their case.

A Proposed Solution for Technical Lag. One feature in the npm environment we would like to see removed is the ability to constrain the specific version of a dependency by default in the package.json file [41]. We believe that there should be a special keyword in package.json, such as *xxx*, for the npm to check the most updated version in its registry and install it automatically to mitigate the issue of technical lag. The keyword *npm outdated* may be unknown to some developers, or the additional workload it would add can be a deterrent factor. However, the keyword that indicates the most recent version should be a default or auto-filled by the package manager to avoid the user interfering with versions. However, this also presents the danger of an update breaking existing systems. The user should always have the option to fix the dependency to a specific version as some releases will provide long-term support or better stability. Thus, the user can have the freedom to decide if the security or functionality of the project is a priority.

Frameworks have also been developed to measure how outdated a system is, balancing security and functionality [14,40]. Calculating a project's recentness score may prove difficult, as each project will have its priorities, such as performance or stability [14]. Gonzalez et al. name this priority as "gold standard" and defines the gold standard to build a lag function that can output the recentness score [14]. For instance, if the user chooses his gold standard as security, the lag function can focus on the number of new updates related to security [14]. If the gold standard is functionality, the lag function can take the number of new features into account [14]. Gonzales et al. even propose other possible lag functions, such as one that simply measures the differences in lines of source code between the most recent and currently deployed dependency or the number of commits of difference between them [14].

To partially implement the framework in npm or PyPi registries, they can mandate reputable libraries to list and classify each update done with the commit. Users can specify their "gold standard" with a specific command after the creation of their projects. npm can use this gold standard to compare the score of the current dependencies with their most recent versions. Now, the user will know his dependencies' recentness in terms of his priorities. Thus, the user can make more informed decisions on whether he should update.

Registries do not have to undergo significant changes to calculate the current projects' scores, as npm can simply take their score as 0. If all future versions require the contributor to classify new features or bug fixes manually, we believe it will not be challenging to implement the new system. Currently, we do not expect to calculate the significance of each particular update, but the user can be asked to assign a score of significance to each bug fix or new feature. We expect several users to avoid spending time with the classification of new features; however, npm can provide incentives to encourage users to participate in the program, such as with extra perks or trust scores.

Automatic Eviction of Squatters. The brute force solution of this problem can be a function that compares the name similarity of a newly published package to an already existing library to determine suspicious packages. Any new package upload can trigger the npm registry to invoke a script that compares the uploaded library name to every other package name in the registry. Suppose the similarity result passes a certain threshold. In that case, the registry can automatically reject the uploaded package, informing the publisher that the package name is too similar to an existing one, and he should choose a new name. However, this approach may prove to be insufficient as the script may not be able to capture every possible scenario.

A simple countermeasure in the literature is the tool TypoGard. It implements a novel detection technique, which uses both the lexical similarity of the package names and the package popularity [29]. It flags certain cases where a user would download an unknown package with a similar name to a popular one before the library can be installed [29]. The tool even caught previously unknown attack attempts, including a package named *loadsh* that imitates the popular npm package *lodash* [30]. An advantage of TypoGard is that it can simply run on the client-side and requires no cooperation from repositories [29]. Researchers evaluated TypoGard on npm, PyPi, and RubyGems, and it has achieved an accuracy of 99.4% with flagging cases [29]. The researchers also measured the overhead of TypoGard and determined it is only 2.5% of the install time [29]. We found that TypoGard is also known as SpellBound in the literature [30].

Squatting also exists in DNS servers, and researchers proposed ML-based solutions to identify suspected squatters in the past [18]. One study proposes approaches based on supervised learning. It specifically uses an ensemble learning classifier that uses various algorithms, such as SVM, that reach an accuracy rate of 88.4% [18]. The study also uses unsupervised learning to identify suspicious domain names by implementing K-means clustering based on specific futures extracted from domain names [18]. Here, the key takeaway seems to be the features chosen to reach this accuracy, such as the number of unique letters or unique numbers in a domain name [18]. For the OS registries case, developers tasked with implementing countermeasures can examine past squatting attacks to create potential features to identify squatters. The features are mostly derived initially by making assumptions of what commonalities of past squatter packages have. These features can even be created by guessing, then feeding them to the ML algorithm to observe which produces the highest accuracy rates.

5.2   Future Direction

First and foremost, all of our suggested countermeasures are theoretical as none of them has been tested in a real-world scenario. We can do additional work by reaching out to the OS registries' admin teams to conduct interviews on the feasibility of our solutions. The admin teams can also have the chance to voice their opinions and common issues they encounter. Based on the issues OS registries face, we may modify some of our solutions or search for additional tools in the literature to meet the admin team's requirements.

To apply ML in the real world for malicious package detection, we need to ensure that the ML model is robust enough against adversarial examples. Adversarial examples are carefully crafted samples to fool the ML model. Previous work has already illustrated how a clever attacker can augment the correctly classified malware with malicious components to cause the ML model to misclassify it [8, 15, 16, 28]. These attacks mostly inject malicious components to the non-reachable parts of the malware, such as padding bytes to the end of the malware or injecting malicious payload to the malware's DOS header. In this way, they can ensure that the original malware purpose is still maintained. However, these attacks can be filtered easily by embedding a code cleaning process into the malicious package detector to filter out non-reachable codes. A more powerful attack can also inject malicious components into the code section [28]. These attacks are more difficult to detect, and further study is required to develop defenses against them.

A useful approach to ease the burden of manual code reviewing for uploaded packages is to measure their similarity to the previously reviewed packages. Suppose the uploaded package is similar to a known malicious package. In that case, it might raise a warning immediately, which can help the reviewers decide whether to reject it into the registry. ML approaches are well suited for detecting code similarity. Previous work provides several ML approaches to detect code similarity in binary-level [17, 27, 37] or decompiled code level [32]. Existing use cases of the proposed methods focus more on detecting software plagiarism and IP theft. Other use cases such as detecting malicious packages are still worth exploring.

Another possible future direction is a project focused on the cost-benefit analysis of using OS package registries. Different organizations with various sizes, fields, or interests may have outweighing benefits or costs using the OS ecosystem, and it is worthwhile to examine the analysis. The aspects of OS security do not seem to be either white or black, and sometimes the risks may justify the benefits for a particular use case.

Future work can also extend the survey to other OS tools aside from public registries. Several companies create internal versions of OS projects to avoid potential hazards. However, these internal versions are closed source, and the OS community cannot benefit from additional features implemented within the firm. Informing the academic society about the security hazards of other OS tools and providing risk mitigation strategies may attract businesses to adopt and actively maintain these projects, thus benefiting the OS community.

6 Conclusion

It seems that research cannot completely remove the risks of npm and PyPi as OS projects mainly rely on volunteer humans and public access. It would be justifiable to assume that these factors constantly create new attack surfaces. The OS registries are open to the public and maintained by humans; however, risk mitigation strategies can be implemented, such as requiring the contributors to use MFA or implementing a trust-score system. It is essentially a trade-off between convenience and security. Nevertheless, the possible opportunity cost of the lack of security must be deliberated carefully before reaching a decision.

One must consider the needs of every individual and establishment when developing software. Certain companies cannot risk a possibility of a data breach as their reputation loss in monetary terms would be much larger than maintaining an army of software developers to convert OS tools to closed source. The lack of trust in the registries would deprive the OS community of the potential knowledge the employees of these firms can bring into the field. A concrete example of this issue is that the Department of Defense is concerned about certain dependencies of their projects being out of their control. This concern would justifiably cause them to refrain from using OS platforms [11]. With countermeasures born from OS security research, admin teams can minimize the risk. The risk can be minimal to the point where some high-security clearance companies may use and even contribute to certain OS packages. Thus, they will strengthen the OS society even further and make software accessible to anyone.

References

1. Abdalkareem, R., Nourry, O., Wehaibi, S., Mujahid, S., Shihab, E.: Why do developers use trivial packages? An empirical case study on npm. In: Proceedings of the 2017 11th Joint Meeting on Foundations of Software Engineering, pp. 385–395 (2017)
2. Abdalkareem, R., Oda, V., Mujahid, S., Shihab, E.: On the impact of using trivial packages: an empirical case study on npm and PyPI. Empir. Softw. Eng. **25**(2), 1168–1204 (2020). https://doi.org/10.1007/s10664-019-09792-9
3. Abuhamad, M., AbuHmed, T., Mohaisen, A., Nyang, D.: Large-scale and language-oblivious code authorship identification. In: Proceedings of the 2018 ACM SIGSAC Conference on Computer and Communications Security, pp. 101–114 (2018)
4. Alfadel, M., Costa, D.E., Shihab, E.: Empirical analysis of security vulnerabilities in python packages. In: 2021 IEEE International Conference on Software Analysis, Evolution and Reengineering (SANER), pp. 446–457. IEEE (2021)
5. Chen, X., Abdalkareem, R., Mujahid, S., Shihab, E., Xia, X.: Helping or not helping? Why and how trivial packages impact the npm ecosystem. Empir. Softw. Eng. **26**(2), 1–24 (2021). https://doi.org/10.1007/s10664-020-09904-w
6. Decan, A., Mens, T., Constantinou, E.: On the evolution of technical lag in the npm package dependency network. In: 2018 IEEE International Conference on Software Maintenance and Evolution (ICSME), pp. 404–414. IEEE (2018)
7. Decan, A., Mens, T., Constantinou, E.: On the impact of security vulnerabilities in the npm package dependency network. In: Proceedings of the 15th International Conference on Mining Software Repositories, pp. 181–191 (2018)
8. Demetrio, L., Coull, S.E., Biggio, B., Lagorio, G., Armando, A., Roli, F.: Adversarial exemples: a survey and experimental evaluation of practical attacks on machine learning for windows malware detection. arXiv preprint arXiv:2008.07125 (2020)
9. Duan, R., Alrawi, O., Kasturi, R.P., Elder, R., Saltaformaggio, B., Lee, W.: Measuring and preventing supply chain attacks on package managers. arXiv preprint arXiv:2002.01139 (2020)
10. Duan, R., Alrawi, O., Kasturi, R.P., Elder, R., Saltaformaggio, B., Lee, W.: Towards measuring supply chain attacks on package managers for interpreted languages. In: NDSS (2021)
11. Ellison, R.J., Goodenough, J.B., Weinstock, C.B., Woody, C.: Evaluating and mitigating software supply chain security risks. Technical report, CARNEGIE-MELLON UNIV PITTSBURGH PA SOFTWARE ENGINEERING INST (2010)
12. Gandotra, E., Bansal, D., Sofat, S.: Malware analysis and classification: a survey. J. Inf. Secur. **5**, 56–64 (2014)
13. Garrett, K., Ferreira, G., Jia, L., Sunshine, J., Kästner, C.: Detecting suspicious package updates. In: 2019 IEEE/ACM 41st International Conference on Software Engineering: New Ideas and Emerging Results (ICSE-NIER), pp. 13–16. IEEE (2019)

14. Gonzalez-Barahona, J.M., Sherwood, P., Robles, G., Izquierdo, D.: Technical lag in software compilations: measuring how outdated a software deployment is. In: Balaguer, F., Di Cosmo, R., Garrido, A., Kon, F., Robles, G., Zacchiroli, S. (eds.) OSS 2017. IAICT, vol. 496, pp. 182–192. Springer, Cham (2017). https://doi.org/ 10.1007/978-3-319-57735-7_17

15. Khormali, A., Abusnaina, A., Chen, S., Nyang, D., Mohaisen, A.: Copycat: practical adversarial attacks on visualization-based malware detection. arXiv preprint arXiv:1909.09735 (2019)

16. Kolosnjaji, B., et al.: Adversarial malware binaries: evading deep learning for malware detection in executables. In: 2018 26th European Signal Processing Conference (EUSIPCO), pp. 533–537. IEEE (2018)

17. Marastoni, N., Giacobazzi, R., Dalla Preda, M.: A deep learning approach to program similarity. In: Proceedings of the 1st International Workshop on Machine Learning and Software Engineering in Symbiosis, pp. 26–35 (2018)

18. Moubayed, A., Injadat, M., Shami, A., Lutfiyya, H.: DNS typo-squatting domain detection: a data analytics & machine learning based approach. In: 2018 IEEE Global Communications Conference (GLOBECOM), pp. 1–7. IEEE (2018)

19. Nataraj, L., Karthikeyan, S., Jacob, G., Manjunath, B.S.: Malware images: visualization and automatic classification. In: Proceedings of the 8th International Symposium on Visualization for Cyber Security, pp. 1–7 (2011)

20. Ni, S., Qian, Q., Zhang, R.: Malware identification using visualization images and deep learning. Comput. Secur. **77**, 871–885 (2018)

21. Ohm, M., Plate, H., Sykosch, A., Meier, M.: Backstabber's knife collection: a review of open source software supply chain attacks. In: Maurice, C., Bilge, L., Stringhini, G., Neves, N. (eds.) DIMVA 2020. LNCS, vol. 12223, pp. 23–43. Springer, Cham (2020). https://doi.org/10.1007/978-3-030-52683-2_2

22. Ohm, M., Sykosch, A., Meier, M.: Towards detection of software supply chain attacks by forensic artifacts. In: Proceedings of the 15th International Conference on Availability, Reliability and Security, pp. 1–6 (2020)

23. Rieck, K., Trinius, P., Willems, C., Holz, T.: Automatic analysis of malware behavior using machine learning. J. Comput. Secur. **19**(4), 639–668 (2011)

24. Santos, I., Devesa, J., Brezo, F., Nieves, J., Bringas, P.G.: OPEM: a static-dynamic approach for machine-learning-based malware detection. In: Herrero, Á., et al. (eds.) International Joint Conference CISIS'12-ICEUTE'12-SOCO'12 Special Sessions. Advances in Intelligent Systems and Computing, vol. 189, pp. 271–280. Springer, Heidelberg (2013). https://doi.org/10.1007/978-3-642-33018-6_28

25. Schultz, M.G., Eskin, E., Zadok, F., Stolfo, S.J.: Data mining methods for detection of new malicious executables. In: Proceedings 2001 IEEE Symposium on Security and Privacy. S&P 2001, pp. 38–49. IEEE (2000)

26. Scott, J.: Signature based malware detection is dead. Institute for Critical Infrastructure Technology (2017)

27. Shalev, N., Partush, N.: Binary similarity detection using machine learning. In: Proceedings of the 13th Workshop on Programming Languages and Analysis for Security, pp. 42–47 (2018)

28. Sharif, M., Lucas, K., Bauer, L., Reiter, M.K., Shintre, S.: Optimization-guided binary diversification to mislead neural networks for malware detection. arXiv preprint arXiv:1912.09064 (2019)

29. Taylor, M., Vaidya, R., Davidson, D., De Carli, L., Rastogi, V.: Defending against package typosquatting. In: Kutyłowski, M., Zhang, J., Chen, C. (eds.) NSS 2020. LNCS, vol. 12570, pp. 112–131. Springer, Cham (2020). https://doi.org/10.1007/ 978-3-030-65745-1_7

30. Taylor, M., Vaidya, R.K., Davidson, D., De Carli, L., Rastogi, V.: Spellbound: defending against package typosquatting. arXiv preprint arXiv:2003.03471 (2020)
31. Ucci, D., Aniello, L., Baldoni, R.: Survey of machine learning techniques for malware analysis. Comput. Secur. **81**, 123–147 (2019)
32. Ullah, F., Wang, J., Farhan, M., Habib, M., Khalid, S.: Software plagiarism detection in multiprogramming languages using machine learning approach. Concurr. Comput. Pract. Exp. **33**(4), e5000 (2021)
33. Vaidya, R.K., De Carli, L., Davidson, D., Rastogi, V.: Security issues in language-based sofware ecosystems. arXiv preprint arXiv:1903.02613 (2019)
34. Vu, D.L., Pashchenko, I., Massacci, F., Plate, H., Sabetta, A.: Towards using source code repositories to identify software supply chain attacks. In: Proceedings of the 2020 ACM SIGSAC Conference on Computer and Communications Security, pp. 2093–2095 (2020)
35. Vu, D.L., Pashchenko, I., Massacci, F., Plate, H., Sabetta, A.: Typosquatting and combosquatting attacks on the python ecosystem. In: 2020 IEEE European Symposium on Security and Privacy Workshops (EuroS&PW), pp. 509–514. IEEE (2020)
36. Wang, H., Ma, C., Zhou, L.: A brief review of machine learning and its application. In: 2009 International Conference on Information Engineering and Computer Science, pp. 1–4. IEEE (2009)
37. Wang, S., Wu, D.: In-memory fuzzing for binary code similarity analysis. In: 2017 32nd IEEE/ACM International Conference on Automated Software Engineering (ASE), pp. 319–330. IEEE (2017)
38. Zapata, R.E., Kula, R.G., Chinthanet, B., Ishio, T., Matsumoto, K., Ihara, A.: Towards smoother library migrations: a look at vulnerable dependency migrations at function level for npm JavaScript packages. In: 2018 IEEE International Conference on Software Maintenance and Evolution (ICSME), pp. 559–563. IEEE (2018)
39. Zerouali, A., Cosentino, V., Mens, T., Robles, G., Gonzalez-Barahona, J.M.: On the impact of outdated and vulnerable JavaScript packages in docker images. In: 2019 IEEE 26th International Conference on Software Analysis, Evolution and Reengineering (SANER), pp. 619–623. IEEE (2019)
40. Zerouali, A., Mens, T., Gonzalez-Barahona, J., Decan, A., Constantinou, E., Robles, G.: A formal framework for measuring technical lag in component repositories-and its application to npm. J. Softw.: Evol. Process **31**(8), e2157 (2019)
41. Zimmermann, M., Staicu, C.A., Tenny, C., Pradel, M.: Small world with high risks: a study of security threats in the npm ecosystem. In: 28th {USENIX} Security Symposium ({USENIX} Security 19), pp. 995–1010 (2019)

Author Index

Printed in the United States
by Baker & Taylor Publisher Services